# Computational Systems Biology Approaches in Cancer Research

# Chapman & Hall/CRC Mathematical and Computational Biology

*Series Editors*
*Xihong Lin*

*Mona Singh*

*N. F. Britton*

*Anna Tramontano*

*Maria Victoria Schneider*

*Nicola Mulder*

This series aims to capture new developments and summarize what is known over the entire spectrum of mathematical and computational biology and medicine. It seeks to encourage the integration of mathematical, statistical and computational methods into biology by publishing a broad range of textbooks, reference works and handbooks. The titles included in the series are meant to appeal to students, researchers and professionals in the mathematical, statistical and computational sciences and fundamental biology and bioengineering, as well as interdisciplinary researchers involved in the field. The inclusion of concrete examples and applications and programming techniques and examples is highly encouraged.

**Introduction to Proteins**
Structure, Function, and Motion, Second Edition
*Amit Kessel, Nir Ben-Tal*

**Big Data in Omics and Imaging**
Integrated Analysis and Causal Inference
*Momiao Xiong*

**Computational Blood Cell Mechanics**
Road Towards Models and Biomedical Applications
*Ivan Cimrak, Iveta Jancigova*

**An Introduction to Systems Biology**
Design Principles of Biological Circuits, Second Edition
*Uri Alon*

**Computational Biology**
A Statistical Mechanics Perspective, Second Edition
*Ralf Blossey*

**Computational Systems Biology Approaches in Cancer Research**
*Inna Kuperstein and Emmanuel Barillot*

For more information about this series please visit: https://www.crcpress.com/Chapman--H allCRC-Mathematical-and-Computational-Biology/book-series/CHMTHCOMBIO

# Computational Systems Biology Approaches in Cancer Research

Edited by
Inna Kuperstein and Emmanuel Barillot

CRC Press
Taylor & Francis Group
Boca Raton London New York

CRC Press is an imprint of the
Taylor & Francis Group, an **informa** business
A CHAPMAN & HALL BOOK

CRC Press
Taylor & Francis Group
6000 Broken Sound Parkway NW, Suite 300
Boca Raton, FL 33487-2742

© 2020 by Taylor & Francis Group, LLC
CRC Press is an imprint of Taylor & Francis Group, an Informa business

No claim to original U.S. Government works

International Standard Book Number-13: 978-0-367-34421-4 (Hardback)

**Visit the Taylor & Francis Web site at**
**http://www.taylorandfrancis.com**

**and the CRC Press Web site at**
**http://www.crcpress.com**

# Contents

# Cancer

## *A New Old Story in the Era of Big Data*

INITIATION AND PROGRESSION OF cancer involve multiple molecular mechanisms, collectively known as hallmarks of cancer. Obviously, we have a lot to discover for improving prognosis and treatment of cancer. The diversity across tumours from different patients and among cells from a single tumour reveals the implication of highly complex mechanisms in the process of tumourigenesis. This complexity confirms that the fundamental aim to find a common mechanism for therapeutic targeting of cancer is not possible. Therefore, the idea of 'personalized' or 'precision' medicine has been suggested, with the goal of finding a tailored treatment regimen for each patient or group of patients according to the individual genetic background and tumour molecular profiles.

This attempt, although ambitious, is achievable thanks to sufficient molecular characterization of cancers accumulated using high-throughput technologies. However, despite availability of cancer multi-dimensional omics and imaging data, they are not fully exploited to provide the clue on deregulated mechanisms that would guide better patient stratification and choice of targeted cancer treatment.

Increasing amounts and diversity of data accumulated in the cancer field has promoted fast evolution of analytical methods and pipelines. This book exposes an umbrella of modern computational systems biology approaches from knowledge formalization to emerging artificial intelligence (AI) and in particular machine learning (ML) approaches applied to cancer research. The book is largely based on didactic lectures given by the authors during the 1st Course on Computational Systems Biology of Cancer at the Institut Curie, France.

## OBJECTIVES OF THE BOOK

The main objective of this book is to share knowledge and to help the reader to grasp, in a fast and efficient way, the most practical methods that exist in the field. The book introduces a wide range of tools that can be directly applied for research and clinical projects.

The second objective of the book is to promote better integration of computational approaches into biological and clinical laboratories. In particular, the book aims to help non-computational researchers and clinicians to get familiar with up-to-date cancer systems biology methodology and to be ready to use these methods.

The authors hope that this collection of approaches will improve the use and interpretation of omics data that nowadays are accumulated in many biological and medical laboratories.

## FOR READERS OF THE BOOK

The book is oriented towards students and researchers from cross-disciplinary areas looking for computational methods adapted to cancer research. It is also addressed to medical doctors, especially oncologists promoting integration of systems biology approaches in clinics. The book also suits a wider audience that is interested in learning in general how cancer research is supported by advanced systems biology methods. Most methods presented here are not limited to cancer, and have a wider use in molecular biology and clinical studies. Therefore, the book is also of interest for researchers involved in other fields of biology and in human health. Finally, this collection of articles can also play a role of didactic support in advanced undergraduate and early graduate school courses in biology, computer science and in interdisciplinary domains.

## WHAT IS THIS BOOK ABOUT?

The book is written in co-authorship with the leading specialists from different fields in cancer systems biology. It consists of *six thematic chapters* each dedicated to a different domain across computational systems biology approaches in cancer. It is structured with the idea to trigger the reader's interest and to provide references and links for immediate and easy exploitation of suggested methods.

Actually, this book can be seen as a *collection of optimized working scenarios* directly applicable in research and clinical teams. Each article in the book describes in a concise form, the rationale and the concept of one approach, supported by an application example. The access to dedicated interactive websites with source code/package/web applications are provided at the end of each article.

The main topics covered in the book include a presentation of pathways resources, tools to study molecular mechanisms of cancer using signalling networks together with omics and clinical data, studies of tumour heterogeneity and tumour microenvironment using single cell approaches, use of drug sensitivity prediction algorithms, especially with respect to immuno-oncology and checkpoint inhibitors, the identification of biomarkers and cancer drivers, the application of mathematical modelling, ML and AI approaches for image analysis and patient stratification in cancer.

## FROM CHAPTER TO CHAPTER

The basis for understanding cancer requires, as a first step, the gathering of a massive amount of knowledge on the altered molecular mechanisms in a computer-readable form. With this respect, **Chapter 1** presents a general pathway database and two disease-specific pathway databases. The three articles show several application examples of the databases for cancer data visualization and interpretation in the context of molecular pathways.

A lot of effort is invested into the characterization of the key cellular players in the tumour microenvironment (TME), especially across the immune component of TME. Thanks to these discoveries, immuno-oncology and in particular the checkpoint inhibitors development field is becoming a promising area. Therefore, **Chapter 2** is dedicated to TME studies from the systems biology perspective and high dimensional data analysis point of view. **Chapter 2.1** demonstrates an approach to derive immune gene signatures directly from tissue transcriptomics data, whereas **Chapter 2.2** shows an oncoimmunology systemic approach to evaluate immune cell infiltration into the tumour and predict tumour clonal evolution via the immunoediting process. Finally, **Chapter 2.3** discusses immune communication network reconstruction and analysis approach to better understand the interplay between immune cell subsets and tumour cells.

One of the major challenges of big data era is how to handle and make sense out of massive amounts of data and how to render them beneficial for explaining complex diseases such as cancer and its comorbidity with other diseases. **Chapter 3** suggests three different analytical platforms. **Chapter 3.1** presents an open source software platform that supports the visualization and analysis of molecular profiling data in the context of functional interaction networks. **Chapter 3.2** exposes a platform for disease comorbidities studies using clinical and omics data. Finally, **Chapter 3.3** suggests a blind deconvolution approach which was applied to a large corpus of transcriptomic data.

The majority of omics data provides us with a snapshot of either a patient sample in a given disease state, or of a cell in a given experimental condition. However, this is insufficient for understanding the dynamics of the coordinated actions between processes that finally result in a particular state of a biological system. **Chapter 4** deals with this challenge, exposing four different scenarios of mathematical modelling of biological systems in cancer. **Chapter 4.1** addresses the question of balance between different T-helper subtypes in cancer by a qualitative dynamical modelling approach. **Chapter 4.2** demonstrates how mathematical modelling of signalling pathways can lead to a hypothesis about new regulatory layers in cell signalling and suggests intervention points in cancer. **Chapter 4.3** combines two approaches, ML, a powerful means to predict drug response complemented by the logical modelling, to bring mechanistic insights helping to rationalize the choice of cancer drugs. **Chapter 4.4** demonstrates how to tailor logical models to particular biological samples such as patient tumours by taking into account individual omics and clinical data that eventually facilitates the choice of patient-specific drug treatments.

Cancer is a very complex disease implicating multiple cell types, where each single cell can undertake a different evolution trajectory having its own destiny. The new single cell analysis methods allow the ability to address this complexity. **Chapter 5.1** tackles high-dimensional data using the ML/AI approach, deciphering stem cell differentiation at the single cell resolution level. To understand intra-tumour heterogeneity, **Chapter 5.2** suggest an approach for mutation calling from single cell sequencing data in order to infer tumour sample phylogeny.

The ultimate goal of many systems biology approaches is to improve patient stratification, to predict response and to rationalize treatment schemes, making them rather combinatorial and, as far as possible, personalized. Accurate multi-omics profiling of patients, complemented with the corresponding clinical records, allows performing studies at the level of cohorts. This topic is addressed in **Chapter 6.1**, demonstrating an approach to identify the disease subtypes using driver pathway signatures. ML/AI approaches are essential in dealing with a combination of omics, imaging and clinical data. **Chapter 6.2** demonstrates an AI/deep learning approach on prognostic modelling and patient stratification by leveraging gene networks as prior knowledge.

**Chapter 6.3** suggests a statistical learning algorithm to correlate the omics and imaging features to clinical outcome in the evaluation of risk of relapse in cancer. Finally, the large-scale interpretation of histopathological images in cancer and subsequent samples stratification today involves ML approaches for computational phenotyping, as suggested in **Chapter 6.4**.

# Pathway Databases and Network Resources in Cancer

## 1.1 SIGNOR AND DISNOR – CAUSAL INTERACTION NETWORKS FOR DISEASE ANALYSIS

*Livia Perfetto, Prisca Lo Surdo, Alberto Calderone, Marta Iannuccelli, Pablo Porras Millan, Henning Hermjakob, Gianni Cesareni and Luana Licata*

### 1.1.1 Summary

SIGNOR (https://signor.uniroma2.it/) – the Signalling Network Open Resource – is a manually curated biological database that captures, organizes and displays signalling interactions as binary, causal relationships between biological entities having a key role in signal transduction and in disease onset. SIGNOR uses the activity flow model to present signalling interaction data through signed directed graphs, as part of a large network that includes about 20,000 interactions between 5,000 biological entities. SIGNOR offers the possibility to generate networks connecting custom lists of entities or browse a collection of pre-assembled signalling pathways.

SIGNOR data can be exploited to extend the understanding of the molecular mechanisms underlying genetic diseases, such as cancer, by linking, for example, disease associated genes with networks of causal relationships. To this end, we have developed the DISease Network Open Resource (DISNOR), a disease-focused resource that combines gene-disease association (GDA) data annotated in the DisGeNET resource with SIGNOR causal interactions. DISNOR can be freely accessed at https://disnor.uniroma2.it/ where approximately 5,000 disease networks, linking more than 4,000 disease genes, can be explored. For each disease curated in DisGeNET, DISNOR links disease genes through manually annotated causal relationships and offers an intuitive visualization of the inferred 'patho-pathways' at different complexity levels.

## 1.1.2 Introduction

Cells are complex and dynamic systems able to modify their behaviour and their morphology in response to internal or environment-induced cues. At a molecular level, chemical, physical or mechanical stimuli are sensed by receptor proteins, which trigger the propagation and the amplification of the signal through the activation of molecular pathways. Pathways are cascades of enzymatic reactions and physical interactions, often culminating in changes of the gene expression profile of the cell. Such rearrangements determine the response of the system to the stimulus and might include metabolic, structural and enzymatic changes of cell assets.[1] The intricate ensemble of these biological processes is named signal transduction.

Given the importance of signal transduction in determining cell phenotype both in physiological and pathological conditions, obtaining an exhaustive understanding of the molecular mechanisms underlying the stimulus-phenotype relationships and formalizing this knowledge into structured data are major goals of systems biology.[2]

As discussed by Le Novere,[3] the main network types currently used to model molecular and gene networks are the so-called 'activity flows' and 'process description'. Briefly: 'activity flow' networks describe signalling events as chains of binary causal relationships (e.g. Protein A activates/inhibits Protein B), while 'process description' networks depict pathways as causal chains of reactions where input molecules are transformed into output molecules (Figure 1.1A).[3]

Databases such as the non-metabolic parts of the Kyoto Encyclopedia of Genes and Genomes pathway (KEGG PATHWAY), SIGNOR and Signalink 0 use 'activity flows', while the metabolic networks in KEGG,

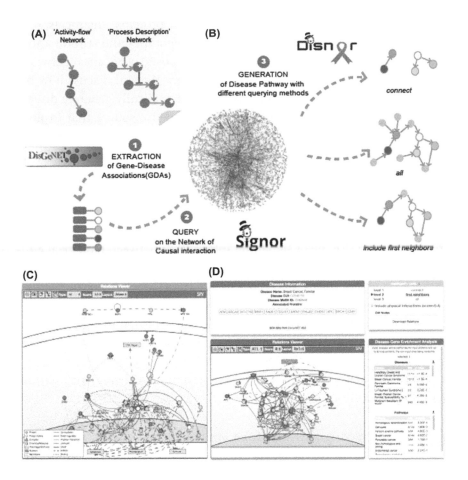

FIGURE 1.1 (A) 'Activity-flow' versus 'Process Description' Networks. (B) Pipeline to produce Disease Networks in DISNOR. (C) Graph extracted from SIGNOR linking gene products that are differentially expressed in BRCA1 versus BRCA2 breast cancer subtypes as shown by Waddell et al.[13] (D) The Familial Breast Cancer as an example of disease network extracted from DISNOR and shown using the graph visualizer (bottom-left). On the top-left box all the information related to the disease is provided. On the top-right part of the panel the querying method (complexity level) chosen is displayed. The bottom-right panel shows the Disease-gene enrichment analysis performed.

Reactome (described in Chapter 1.2) and NaviCell adopt a 'process description' model.[4-8]

In this chapter we will describe the content and the website of the SIGNOR, a biological database that stores manually curated, binary causal interactions between biological entities such as proteins, chemicals, protein families, complexes, small molecules, phenotypes and stimuli annotated from scientific literature. SIGNOR is a repository of causal networks that can support systems medicine approaches. We will discuss how the data provided in SIGNOR can be used to systematically generate disease networks within DISNOR[9], a recently developed resource that exploits SIGNOR to extend our understanding of the molecular mechanisms underlying genetic diseases, such as cancer, by linking disease associated genes via a network of causal relationships.

### 1.1.3 Approach and Application Example

SIGNOR is a freely available open-source database that focuses on the annotation and storage of signalling interactions between biological entities and adopts the 'activity flow' model to represent biological processes.

Each signalling interaction is represented as a directed binary relationship between entities, one playing the role of the regulator and the other being the regulation target. Entities in SIGNOR belong to six categories: proteins, chemicals, small molecules, phenotypes, protein families or complexes.

Every interaction in SIGNOR is manually curated from the literature and annotated with details about:

1) The interacting partners: Each entity is defined by an ID that can either be inherited from an external resource (UniProt IDs for proteins, ComplexPortal IDs for complexes) or defined within SIGNOR, using an internal Controlled Vocabulary.

2) The sign or effect of the interaction: A regulator might up- or downregulate the target by modulating its activity or quantity.

3) The molecular mechanism underlying the regulatory interaction (e.g. phosphorylation, ubiquitination, etc.).

4) Whether the interaction is direct (the regulator entity is immediately upstream of the regulated entity) or not.

5) A reference (usually the PubMedID) and a sentence (extracted from the related PubMedID) supporting the interaction.

Additional details, such as the phosphorylated residue or the cell line/tissue in which the interaction occurs are also captured, whenever possible.

The manually curated data is originally mapped by the curators to *Homo sapiens* protein IDs irrespective of the experimental system that was used to provide the evidence. Orthology mapping is then used to map further the entries to *Mus musculus* and *Rattus norvegicus*.

All data in SIGNOR are annotated following specific curation rules and using controlled vocabularies developed by the SIGNOR team in collaboration with other international signalling resources.

In particular, the PSI-MI controlled vocabulary has been extended by adding additional terms to represent causal interactions. The extended PSI-MI tab-delimited format, also called CausalTab (or PSI-MITAB2.8) is a common standard for the representation and dissemination of signalling information.[10]

SIGNOR can be freely accessed at https.//signor.uniroma2.it/, where it is possible to query the database providing one or multiple entities. In a search for a single entity, all the details and the activity-flow network centred on the query entity are displayed. In a search for multiple entities, the user has the option to display the interactions occurring between the query or seed entities, choosing from three different searching methods: *connect*, *all* and *include first neighbours* (Figure 1.1B).

As shown in Figure 1.1B, the *connect* method results in a network displaying only direct connections between input list, while the *all* method returns any possible interaction found for every entity in the input; finally, the *include first neighbours* method is a multi-step method that initially performs an *all* search, followed by a pruning step that removes all the irrelevant interactions (interactions that do not link input entities).

All the query types mentioned return a graphic view,[11] a schematic and detail-rich representation of the retrieved interactions presented using a graph where nodes are gene products or other biologically relevant entities and edges represent the causal relationships between them (Figure 1.1C). The graphic visualizer is an intuitive, customizable and dynamic display of the interactions. The interpretation of the signalling cascade is facilitated by the adoption of symbols and colour codes for nodes and edges, whose explanation is provided in the graph legend (Figure 1.1C). Also, the graphic viewer contains four cellular compartments: the extracellular space, the plasma membrane, the cytoplasm and the nucleus. Entities are arranged in these compartments according to external annotations, thus offering a readily readable image of how the signal cascade propagates from the receptors and their ligands to the transcription factors.

A short guide explaining all the query methods, data details and the visualization tool is available at Lo Surdo et al.[12] Data are available via REST API for programmatic access and as CausalTab,[10] downloadable from the 'download' section (https://signor.uniroma2.it/downloads.php).

Even though SIGNOR curators currently capture relationships for any possible human gene, the initial goal of the SIGNOR project was the annotation of causal networks whose perturbation cause pan-cancer insurgence and progression. The curation approach was focused on causal relationships of core oncogenes and tumour suppressors and on the annotation of how oncogenes and tumour suppressors are linked to onco-relevant cellular phenotypes (e.g. 'apoptosis', 'cell cycle progression', 'cell proliferation', etc.).

Over the years, SIGNOR curators have also focused their attention on the curation of post translation modification reactions, in particular, more than 9,000 phosphorylation reactions have been annotated that represent a valuable dataset to map and interpret high content phosphoproteomic data.

Data from SIGNOR can be used to understand the mechanism explaining a particular molecular phenotype. For example, Waddell and collaborators[13] identified 393 genes that show a different expression profile in BRCA1 versus BRCA2 familial breast cancer samples. These genes do not significantly enrich any particular pathway or biological process. We mapped these genes to 124 UniProt identifiers and used this list of proteins to query SIGNOR using the *include first neighbour* searching method. As shown in Figure 1.1C, SIGNOR returns a graph of causal interactions connecting 28 proteins from the input list. These interactions are likely to be differentially altered in the two cancer conditions. In addition to that, SIGNOR provides the link between these proteins and phenotypes that are commonly affected in malignancy (e.g. Proliferation, Apoptosis, etc.). The graph provided might offer new insight about the molecular mechanisms underlying the two cancer types.

The ultimate goal of SIGNOR is to support systems medicine approaches. We discuss in the following section how the data provided in SIGNOR are exploited within DISNOR, to systematically generate disease networks. The premise behind DISNOR is that genes associated with the same or similar diseases commonly reside in the same neighbourhood in an interaction network and are observed to affect common biological processes or pathways.[14] As a consequence, the interactions linking genes that are mutated in patients affected by the same disease can be used to reconstruct the molecular pathway explaining the pathological phenotype.

DISNOR links disease-associated genes (extracted from the DisGeNET resource[15]) via causal relationships annotated within the SIGNOR database.

DISNOR can be freely accessed at https://disnor.uniroma2.it/, where it is possible to explore approximately 5,000 disease networks, linking more than 4,000 disease genes (Figure 1.1D). For more than 80% of the malignancies in DisGeNET, DISNOR suggests a graph with an average network size of 20 nodes and 280 edges, covering more than 75% of cancer-associated genes (extracted from DisGeNET). For each disease curated in DisGeNET, DISNOR links disease genes by manually annotated causal relationships and offers an intuitive visualization of the inferred 'patho-pathways' at different complexity levels. An example of that is provided in Figure 1.1D, showing the graph obtained for the Familial Breast Cancer.

User-defined gene lists are also accepted in the query pipeline. For queries containing lists of genes – either annotated within DisGeNET or user-defined – DISNOR allows the user to submit a gene set enrichment analysis on KEGG-defined pathways or on the lists of proteins associated with the inferred disease pathways. This function offers additional information on disease-associated cellular pathways and disease similarity.

## 1.1.4 Discussion and Perspectives

Network modelling is a powerful tool for a coherent representation of the complexity of biological systems in a simple and unified model and can provide further knowledge on a biological model of interest and help to assess the validity of a hypothesis, or the correctness of a proposed mechanism with crucial implications in cancer research.[16]

SIGNOR is a resource offering a wide range of logic relationships that can be managed and rearranged by experts to assemble an ad hoc prior knowledge network.

SIGNOR is a directed graph of regulatory relationships, comprising approximately 5,000 nodes and 20,000 edges; the data in this resource has already been reported to support network-based integration approaches in personalized medicine studies.[17,18] The SIGNOR dataset has also been used in multi 'omics' and computational modelling studies to better understand cancer dynamics and to discover new, effective therapeutic approaches or predict compound toxicity.[19,20]

Similarly, Sacco et al.[21] have recently reviewed computational approaches exploiting the more than 9,000 phosphorylation sites annotated in SIGNOR to improve extraction of signalling information from phosphoproteomic datasets.[21]

Future goals are to obtain a higher coverage of other PTMs, such as ubiquitination and acetylation, involved in many fundamental cellular processes and in the onset of several diseases.

## Useful Resources

Website: https://signor.uniroma2.it/

Website: https://disnor.uniroma2.it/

User Guide: https://signor.uniroma2.it/user_guide.php

## References

1. Lee, M.J. & Yaffe, M.B. Protein regulation in signal transduction. *Cold Spring Harb. Perspect. Biol.* **8**, (2016).
2. Barabási, A.L. & Oltvai, Z.N. Network biology: Understanding the cell's functional organization. *Nat. Rev. Genet.* **5**, 101–113 (2004).
3. Le, N.N. Quantitative and logic modelling of molecular and gene networks. *Nat. Rev. Genet.* **16**, 146–158 (2015).
4. Kanehisa, M., Furumichi, M., Tanabe, M., Sato, Y. & Morishima, K. KEGG: New perspectives on genomes, pathways, diseases and drugs. *Nucleic Acids Res.* **45**, D353–D361 (2017).
5. Perfetto, L. et al. SIGNOR: A database of causal relationships between biological entities. *Nucleic Acids Res.* **44**, D548–54 (2016).
6. Fazekas, D. et al. SignaLink 2 – A signalling pathway resource with multi-layered regulatory networks. *BMC Syst. Biol.* **7**, 7 (2013).
7. Fabregat, A. et al. The reactome pathway knowledgebase. *Nucleic Acids Res.* **46**, D649–D655 (2018).
8. Kuperstein, I. et al. NaviCell: A web-based environment for navigation, curation and maintenance of large molecular interaction maps. *BMC Syst. Biol.* **7**, 100 (2013).
9. Lo, P.S. et al. DISNOR: A disease network open resource. *Nucleic Acids Res.* **46**, D527–D534 (2018).
10. Perfetto, L. et al. CausalTAB: The PSI-MITAB 2.8 updated format for signalling data representation and dissemination. *Bioinformatics* (2019). doi:10.1093/bioinformatics/btz132.
11. Calderone, A. & Cesareni, G. SPV: A javascript signalling pathway visualizer. *Bioinformatics* **34**, 2684–2686 (2018).

12. Lo, P.S., Calderone, A., Cesareni, G. & Perfetto, L. SIGNOR: A database of causal relationships between biological entities-a short guide to searching and browsing. *Curr. Protoc. Bioinformatics* **58**, 8.23.1–8.23.16 (2017).

13. Waddell, N. et al. Subtypes of familial breast tumours revealed by expression and copy number profiling. *Breast Cancer Res. Treat.* **123**, 661–677 (2010).

14. Menche, J. et al. Integrating personalized gene expression profiles into predictive disease-associated gene pools. *NPJ Syst. Biol. Appl.* **3**, 10 (2017).

15. Piñero, J. et al. DisGeNET: A comprehensive platform integrating information on human disease-associated genes and variants. *Nucleic Acids Res.* **45**, D833–D839 (2017).

16. Kestler, H.A., Wawra, C., Kracher, B. & Kühl, M. Network modeling of signal transduction: Establishing the global view. *BioEssays* **30**, 1110–1125 (2008).

17. Dimitrakopoulos, C. et al. Network based integration of multi omics data for prioritizing cancer genes. *Bioinformatics* **34**, 2441–2448 (2018).

18. Luo, W.M., Wang, Z.Y. & Zhang, X. Identification of four differentially methylated genes as prognostic signatures for stage I lung adenocarcinoma. *Cancer Cell Int.* **18**, 60 (2018).

19. Kanhaiya, K., Czeizler, E., Gratie, C. & Petre, I. Controlling directed protein interaction networks in cancer. *Sci. Rep.* **7**, 10327–10327 (2017).

20. Alexander-Dann, B. et al. Developments in toxicogenomics: Understanding and predicting compound-induced toxicity from gene expression data. *Mol. Omics* **14**, 218–236 (2018).

21. Sacco, F., Perfetto, L. & Cesareni, G. Combining phosphoproteomics datasets and literature information to reveal the functional connections in a cell phosphorylation network. *Proteomics* **18**, e1700311 (2018).

## 1.2 REACTOME: A FREE AND RELIABLE DATABASE TO ANALYZE BIOLOGICAL PATHWAYS

*Thawfeek Varusai, Steven Jupe, Lisa Matthews, Konstantinos Sidiropoulos, Marc Gillespie, Phani Garapati, Robin Haw, Bijay Jassal, Florian Korninger, Bruce May, Marija Milacic, Corina Duenas Roca, Antonio Fabregat, Karen Rothfels, Cristoffer Sevilla, Veronica Shamovsky, Solomon Shorser, Guilherme Viteri, Joel Weiser, Guanming Wu, Lincoln Stein, Peter D'Eustachio, Henning Hermjakob*

### 1.2.1 Summary

Reactome is an open-source, open access, manually curated and peer-reviewed human biomolecular pathway database providing intuitive bioinformatics tools for the visualization, interpretation and analysis of pathway knowledge to support basic and clinical research, genome analysis, modelling, systems biology and education. Biological processes are represented as interconnected molecular events or 'reactions' in Reactome. Reactions are the 'steps' in pathways and can be any molecular event in biology. Pathways are organized hierarchically and often have sub-pathways. At the highest hierarchical level, pathways are represented in a genome-wide overview. At intermediate and lower levels, they are represented as interactive textbook-style illustrations and detailed pathway diagrams. Pathways in several other species are computationally inferred from the manually curated human pathways. Reactome includes tools for analysing the pathway context of user data, mapping expression or other quantitative data onto pathways or extending pathways with interactions from external databases. Reactome, therefore, provides biological pathway information in an efficient manner and tools to analyze this data, which would not be possible with traditional literature-style data.

### 1.2.2 Introduction

Modern, high-throughput studies in biology provide abundant and complex data, posing a huge challenge for analysis and interpretation. At the same time, the available biomolecular knowledge rises rapidly; a simple PubMed query for 'Pathway' returns about 50,000 publications per publication year. It is an ongoing challenge for researchers to keep up-to-date with research developments in their fields and identify relevant research to support their own studies without devoting too much time collecting unconnected information. What is desired is an open platform that hosts reliable information on biological data in a format that is easy to use.

Typically, researchers rely on scientific literature for different aspects of biological information. One of the most sought-after information types are mechanistic details of cellular reactions and pathways. Pathway information is often presented in scientific literature as figures or diagrams. For the human reader, graphics are an intuitive way to visualize a biological process, but there are several limitations to these formats. For example:

- The information is spread across many thousands of papers.

- It is not accessible for computational reuse.

- It is not possible to use the pathways interactively.

There is a need to address these basic challenges in the field.

Founded in 2003, Reactome is a freely available, open-source database of signalling and metabolic molecules and their relations organized into biological pathways and processes.[1] Reactome events are based on the peer-reviewed literature, extracted from papers by biologists who are experts in their field, assisted by scientific curators who add consistent structure to the information and represent it in the database. This process verifies the details and provides a consistent structure to the data, making it accessible for computational data mining and re-use. Reactome is therefore a curated, quality-assured resource of pathway information, available as diagrams and descriptions on the Reactome website and downloadable in standard formats such as BioPax[2] and SBML.[3]

Reactome is designed to literally give the user a graphical map of known biological processes and pathways that is also an interface, which the user can 'click through' to authoritative detailed information on components and their relations.[4] The website also provides integrated tools for analysis of user data, for species comparison and the extension of pathways with protein–protein and protein–compound interaction data.[5] The Reactome database and website enable scientists, researchers, students and educators to find, organize and utilize biological information to support data visualization, integration and analysis.

## 1.2.3 Approach and Application Example

Reactome is an extremely useful starting point for (1) identifying complete molecular details of a pathway and (2) learning the pathways, events and complexes that include a protein, simple chemical compound or set of either or both of these. At present, Reactome's *Analyse Data* tool allows analysis of either lists of genes, proteins and/or compounds or lists with associated quantitative data such as expression level, concentration or frequency. The features and tools in Reactome cater to the needs of several fields – clinicians, geneticists, genomics researchers and molecular biologists trying to interpret the results of high-throughput experimental studies; bioinformaticians seeking to develop novel algorithms for mining knowledge from genomic studies; and systems biologists aiming to build predictive models of normal and disease variant pathways.

We shall now demonstrate the pathway enrichment analysis tool in Reactome using the familial breast cancer example from Chapter 1.1. The DISNOR[6] website provides network information on genes associated with familial breast cancer and the first neighbour interactors of these genes using data from DisGeNET[7] and SIGNOR.[8] We will use this list of associated genes and their first neighbour interactors to identify enriched pathways in Reactome.

UniProt[9] IDs of the breast-cancer-associated genes and their first neighbour interactors are listed in a single column with a header that represents the data. Reactome requires this header name to start with a # symbol and be devoid of blank spaces. This list is uploaded or pasted in the analysis tool and pathway enrichment is run. The resulting page shows three panels – a hierarchy panel on the left, a details panel at the bottom and a diagram panel in the middle. The three panels are interlinked with each other and can be used interactively.

The diagram panel displays graphical representations of pathways at different levels of detail and complexity (Figure 1.2). The genome-wide overview diagram shows all the processes in Reactome as concentric rings with each hub representing a process and nodes representing sub-pathways. Upon pathway enrichment analysis, significantly expressed pathways are highlighted in the overview diagram (Figure 1.2A). Clicking on a pathway node will show more details about the pathway in an easy-to-understand textbook-style diagram. In a pathway enrichment analysis, the degree of overlap between user data and pathway entities is indicated in this diagram in the form of a highlighted progress bar (Figure 1.2B). Double-clicking a sub-pathway from the textbook-style diagram will display the pathway diagram that contains details of the reactions and proteins in this pathway. In a pathway enrichment analysis, matching entities between user data and Reactome are shown as fully and partially

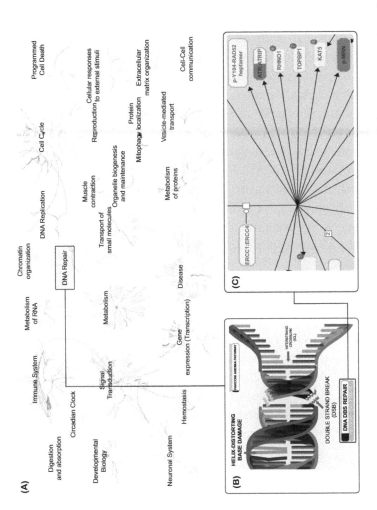

FIGURE 1.2 Pathway enrichment analysis of genes associated with familial breast cancer. (A) Process overview of enriched pathways highlighting nodes that are significantly expressed in the user data. (B) Textbook-style figure of the most significantly enriched pathway – DNA repair process. The degree of similarity between user data and pathway is represented in the form of a progress bar in the figure. (C) Pathway diagram of homologous recombination of DNA within DNA repair process showing the various proteins, complexes and reactions. Proteins and complexes are coloured fully or partially depending on matching user data.

coloured proteins and complexes in the pathway diagram (Figure 1.2C). This approach of visualizing pathways at different levels of complexity in Reactome allows users to interactively understand the underlying biology of their data.

The details panel at the bottom contains elaborate information on the selected event or pathway including manually authored descriptive, structural and expression information. In an enrichment analysis, the details panel contains statistical information of the analysis using user data. Pathway over-representation analysis is a statistical (hypergeometric distribution) test that determines whether certain Reactome pathways are over-represented (enriched) in the submitted data. It answers the question 'Does my list contain more proteins for pathway X than would be expected by chance?' This test produces a probability score, which is corrected for false discovery rate using the Benjamini-Hochberg method.[10] The resulting enriched pathways are ranked in ascending order of False Discovery Rate, which means the most statistically significant pathways are shown first. The hierarchy panel on the left contains the structured tree of pathways in Reactome with parent and child pathways. In a pathway enrichment analysis, the hierarchy panel additionally displays statistical metrics against each pathway.

The three panels are interconnected and pathway enrichment analysis results can be explored by interacting with different elements in each panel. In our example of genes associated with familial breast cancer, an enrichment analysis lists all the pathways in Reactome with statistically significant pathways ranked at the top. One way to analyze this data is to check if top-ranked pathways belong to a certain physiological process. We observe that several of the over-represented sub-pathways such as synthesis-dependent strand annealing, DNA double strand break repair and homology-directed repair are grouped under DNA repair process.

Reactome analysis suggests that the DNA repair process is highly involved in familial breast cancer (Table 1.1). We can try to validate this finding by crosschecking with other sources. Several studies report the contribution of DNA repair process to familial breast cancer.[11] Our analysis corroborates with data in literature. Furthermore, pathway enrichment analysis performed on breast-cancer-associated genes alone (from DisGeNET; result not shown) or with their first neighbour interactors (from SIGNOR) give the same result – over-represented DNA repair process pathways. This indicates that Reactome is in line with other manually

TABLE 1.1   Top Five Enriched Pathways Using Genes Associated with Familial Breast Cancer

| Pathway Name | Entities | | | | Reactions | |
|---|---|---|---|---|---|---|
| | Found | Ratio | p-value | FDR[a] | Found | Ratio |
| Resolution of D-loop Structures through SDSA | 10/26 | 0.002 | 1.11E-16 | 2.78E-15 | 1/1 | 8.70E-05 |
| Regulation of TP53 through Phosphorylation | 15/92 | 0.008 | 1.11E-16 | 2.78E-15 | 26/26 | 0.002 |
| DNA Double-Strand Break Repair | 19/148 | 0.013 | 1.11E-16 | 2.78E-15 | 98/107 | 0.009 |
| G2/M DNA Damage Checkpoint | 13/78 | 0.007 | 1.11E-16 | 2.78E-15 | 11/12 | 0.001 |
| Homology Directed Repair | 16/120 | 0.011 | 1.11E-16 | 2.78E-15 | 43/50 | 0.004 |

[a]  False Discovery Rate.

curated resources such as SIGNOR, thereby adding more confidence to the analysis tool.

## 1.2.4  Discussion and Perspectives

Apart from pathway enrichment, Reactome contains other analysis tools and interactive options, which are not discussed here for the sake of brevity. The intention of this chapter is to demonstrate the use of Reactome as a biological pathway database and analysis platform. Detailed information on how to use Reactome is available on the website: https://reactome.org/. Reactome provides excellent guidance to users with extensive documentation on how to navigate, explore and use the various analysis tools. Tutorials are available in various forms including short video tours, webinars, elaborate user guide documents and self-paced online training courses.

Content in Reactome is dynamic and intended to expand and undergo revision as new discoveries are published and become accepted. To complement the work of our full-time professional curators, we encourage community contributions to expedite the inclusion of critical pathway information in the database. The database is updated on a quarterly basis, meaning new content will be available every three months. New features and tools for pathway visualization and analysis are also under constant development to meet the changing needs of the research community. Thus, Reactome is an evolving open-source database aiming to provide a highly reliable service to the scientific community working with biological pathways.

## Useful Resources

Website: https://reactome.org/

User Guide: https://reactome.org/user/guide

Online Training: https://www.ebi.ac.uk/training/online/

Helpdesk: help@reactome.org

## References

1. Fabregat, A. et al. The reactome pathway knowledgebase. *Nucleic Acids Res.* **44**, D481–D487 (2016).
2. Demir, E. et al. The BioPAX community standard for pathway data sharing. *Nat Biotechnol* **28**, 935–942 (2010).
3. Hucka, M. et al. The systems biology markup language (SBML): A medium for representation and exchange of biochemical network models. *Bioinformatics* **19**, 524–531 (2003).
4. Sidiropoulos, K. et al. Reactome enhanced pathway visualization. *Bioinformatics* **33**, 3461–3467 (2017).
5. Fabregat, A. et al. Reactome pathway analysis: A high-performance in-memory approach. *BMC Bioinformatics* **18**, 142 (2017).
6. Lo Surdo, P. et al. DISNOR: A disease network open resource. *Nucleic Acids Res.* **46**, D527–D534 (2018).
7. Pinero, J. et al. DisGeNET: A comprehensive platform integrating information on human disease-associated genes and variants. *Nucleic Acids Res.* **45**, D833–D839 (2017).
8. Perfetto, L. et al. SIGNOR: A database of causal relationships between biological entities. *Nucleic Acids Res.* **44**, D548–D554 (2016).
9. Apweiler, R. et al. UniProt: The universal protein knowledgebase. *Nucleic Acids Res.* **32**, D115–D119 (2004).
10. Hochberg, B. Controlling the false discovery rate: A practical and powerful approach to multiple testing. *J R Stat Soc.* **57** (1995).
11. Girard, E. et al. Familial breast cancer and DNA repair genes: Insights into known and novel susceptibility genes from the GENESIS study, and implications for multigene panel testing. *Int J Cancer.* **144**, 1962–1974 (2018).

## 1.3 ATLAS OF CANCER SIGNALLING NETWORK: AN ENCYCLOPEDIA OF KNOWLEDGE ON CANCER MOLECULAR MECHANISMS

*Jean-Marie Ravel, L. Cristobal Monraz Gomez, Maria Kondratova, Nicolas Sompairac and Inna Kuperstein*

### 1.3.1 Summary

Cancerogenesis is associated with aberrant functioning of a complex network of molecular interactions, simultaneously affecting multiple cellular functions. Therefore, the successful application of bioinformatics and systems biology methods for analysis of high-throughput data in cancer research heavily depends on availability of global and detailed reconstructions of cancer-specific molecular networks amenable for computational analysis. We present here Atlas of Cancer Signalling Network (ACSN), a pathway database and an interactive comprehensive network map of molecular mechanisms deregulated in cancer cells and tumour microenvironment. The resource includes tools for map navigation, visualization and analysis of molecular data in the context of signalling network maps. Constructing and updating ACSN involves careful manual curation of molecular biology literature and participation of experts in the corresponding fields. The cancer-oriented content of ACSN comprehensively covers the hallmarks of cancer and is completely original. This disease-specific resource can be useful to study and interpret molecular perturbations in cancer, among others, explaining drug resistance, suggesting intervention points, finding phenotype shifts through disease progression, explaining susceptibility to a particular type of cancer, studying disease comorbidities and beyond.

### 1.3.2 Introduction

Carcinogenesis represents aberrant functioning of a complex network of molecular interactions and processes, such as cell cycle, regulated cell death, DNA repair and replication, cell motility and adhesion, cell survival mechanisms, immune processes, angiogenesis, tumour microenvironment and many others. These processes are collectively or sequentially involved in tumour formation and modified as the tumour evolves.

The scientific literature often suggests that, in pathological situations, the normal cell signalling network is altered by deregulated coordination between pathways or disruption of existing molecular pathways, rather than by creating completely new signalling pathways. The most common abnormalities in pathological situations are perturbations at the gene expression level, protein abundance or protein posttranslational modifications, irregular 'firing' or silencing of particular signals, wrong subcellular localization of particular molecules and so on. Such quantitative rather than qualitative network changes, compared with the normal cell signalling, could be studied in the context of comprehensive signalling networks by analyzing experimental data obtained from tumour samples, patient-derived xenografts, cancer-related cell lines or animal models. This approach helps to understand the interplay between molecular mechanisms in cancer and to decipher how gene and protein interactions govern the hallmarks of cancer[1] in specific settings.

Despite the existence of a large variety of pathway databases and resources[2], only few of them are cancer-specific, and none of these resources depict the processes with enough granularity. In addition, pathway browsing interfaces are becoming more important for cancer researchers and clinicians; however, further improvements are required.

### 1.3.3 Approach and Application Example

ACSN (https://acsn.curie.fr) is a web-based resource of multi-scale biological maps depicting molecular processes in cancer cell and tumour microenvironment.[3] The core of the Atlas is a set of interconnected cancer-related signalling and metabolic network maps. Molecular mechanisms are depicted on the maps at the level of biochemical interactions, forming a large seamless network of over 8,000 reactions covering close to 3,000 proteins and 800 genes and based on more than 4,500 scientific publications (Figure 1.3A).

Maps are constructed using the Systems Biology Graphical Notation standard with the CellDesigner tool.[4] ACSN maps are reaction networks represented using process diagram description graphical language. The maps are manually created based on the information extracted from scientific literature and contain information about interactions between proteins, protein modifiers, post-translational modifications, protein complexes, genes, RNAs, microRNAs, small molecules and drugs. Each map covers hundreds of molecular players and reactions. Each entity on the map is annotated by references to the article and specific notes added by the map manager. Maps are curated by the specialists in the corresponding fields.[5]

The Atlas is a 'geographic-like' interactive 'world map' of molecular interactions leading the hallmarks of cancer as described by Hanahan

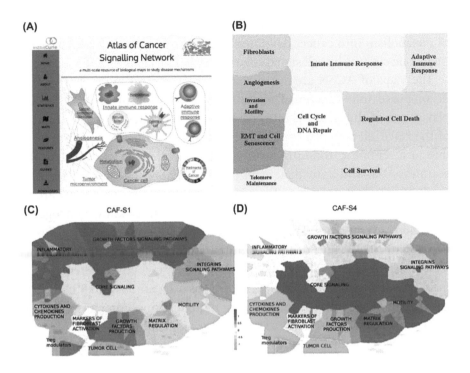

FIGURE 1.3 (A) ACSN webpage provides access to maps, guides and download-able materials. (B) Top-level view of ACSN map demonstrating geographical-like structure of the resource where each territory indicates biological processes in cancer cell and in tumour microenvironment. Visualization of transcriptome profiles for (C) CAF-S1 and (D) CAF-S4 in the context of cell type-specific CAF map. Red-up-regulated and green to down-regulated functional modules.

and Weinberg. The Atlas is created with the use of systems biology standards and amenable for computational analysis. ACSN is composed of 13 comprehensive maps of molecular interactions. There are six maps covering signalling processes involved in cancer cell and four maps describing tumour microenvironment. In addition, there are three cell type-specific maps describing signalling within different cells types frequently surrounding and interacting with cancer cells. This feature of ACSN reflects complexity of tumour microenvironment. The maps of ACSN are interconnected; the regulatory loops within cancer cell and between cancer cell and tumour microenvironment are systematically depicted (Figure 1.3B).

The cross-talk between signalling mechanisms and metabolic processes in the cancer cells is explicitly depicted thanks to a new feature of the Atlas: ACSN is now connected to ReconMap,[6] the largest graphical representation of human metabolism. This allows studying involvement of metabolism into cancer development and understanding regulation of metabolic circuits by cell signalling mechanisms in cancer and vice versa.

The maps of ACSN are organized in a hierarchical manner. They are decomposed into functional modules with meaningful network layout. Navigation of the ACSN is intuitive thanks to Google Maps-like features of the NaviCell web platform.[7] The exploration of the Atlas is simplified due to a semantic zooming feature, allowing the user to visualize the seamless Atlas and individual maps from the ACSN collection at different levels of details description. The top-level view shows the general architecture of the map. The next zoom exposes locations of known oncogenes or tumour suppressors. The third zoom level adds some of the most participating proteins and complexes in the atlas. Furthermore, two zooms visualize components of the atlas and reaction edges between them. At the most detailed zoom level, all details of maps are demonstrated including names of all entities, post-translational modifications, complexes' names, reaction IDs and all regulators and catalyzer of reactions.

Cross-referencing with other databases and links to the scientific papers that were used for creating the Atlas allows the user to study in depth the knowledge represented in ACSN. The ACSN content is permanently expanded with new signalling network maps and updated by the latest discoveries in the field of cancer-related cell signalling. In addition, ACSN is not only a cancer-oriented database: ACSN maps depict fundamental cell signalling processes and can be used in many domains of molecular biology.

The NaviCell web-based data analysis toolbox integrated in ACSN allows importing and visualizing heterogeneous omics data on top of the ACSN2.0 maps and performing standard functional analysis.[8] In addition, NaviCom (https://navicom.curie.fr) is a platform for generating interactive-network-based molecular portraits using high-throughput datasets.[9] NaviCom connects between the cBioPortal database (www. cbioportal.org)[10] and NaviCell web service and allows displaying of various high-throughput data types simultaneously on the network maps in a user-friendly way.[11]

To our knowledge, this is the only resource in the field that systematically gathers together unique and up-to-date information about mechanisms deregulated in cancer cells and tumour microenvironment. The entire content of ACSN2.0 is available on the website and downloadable in various formats. Essential scripts and documentation regarding the maps construction process, implementing maps into the NaviCell platform and the NaviCell visualization tool are provided through the GitHub repository and accessible from the ACSN website, making ACSN compliant with FAIR principles.

We describe here one example of multiple applications,[12] on cell type-specific signalling network maps which help to reveal signatures of cell heterogeneity and polarization in tumour microenvironment.[13] The molecular complexity of tumor microenvironment (TME) creates a bottleneck in interpretation of cancer omics data. To address this challenge, we demonstrate that systematic graphical representation of cell type-specific molecular mechanisms governing polarization of different TME components is an important step in data interpretation. To describe the balance between components of the TME and to analyze impact of non-immune cells as cancer-associated fibroblasts (CAF), information on related molecular mechanisms was systematically collected and represented in a form of a comprehensive network map. The modular map is covering the main functions of CAFs in cancer as interactions with extracellular matrix components, signalling coordinating involvement of CAFs in tumour growth and interactions of CAFs with immune system in TME. The CAF network map contains 11 functional modules, showing mechanisms associated with pro-tumoural CAF activity.

Analysis and interpretation of expression patterns from breast cancer CAFs samples in the context of the network map helped to characterize different CAF subsets with distinct molecular properties. The enrichment analysis of transcriptome using the gene sets from functional modules of the CAF map highlighted that the CAF-S1 subset exhibits a high expression of immune signatures, including cytokines production and modulation of regulatory T lymphocytes (Tregs), promoting an immunosuppressive microenvironment (Figure 1.3C); whereas the CAF-S4 subset is exhibiting matrix regulation and motility mechanisms, indicating that the CAF-S4 subset most probably modulates properties of extracellular matrix and facilitates tumour invasion (Figure 1.3D). The application of this systems biology resource for identification of the CAFs subset in breast cancer has broader context for studying different components of TME[13]).

### 1.3.4 Discussion and Perspectives

There is an ongoing effort to include ACSN content into various open-source platforms such as MINERVA[14] and NDEx, the Network Data Exchange platform.[15] To make the applications of ACSN broader we ensure that the resource is compatible with, and well connected to, downstream analysis pipeline tools in the GARUDA connectivity platform.[16] Further, including ACSN into databases such as Pathway Commons[17] and WikiPathways[18] will allow the community to use this cancer-related resource together with generic pathways.

In addition, integrating two comprehensive manually curated network maps such as ACSN[3] and ReconMap from the Virtual Metabolic Human initiative[6] allows elucidation of the crosstalk between metabolic and signalling pathways. It opens an opportunity to model coordination between signalling pathways and metabolism using various systems biology approaches for multi-level omics data analysis in the context of the biological network maps that allow defining 'hot' areas in molecular mechanisms and point to key regulators in physiological or in pathological situations and beyond.

The ACSN resource and analytical tools described in this chapter can serve as basis for a more general approach to understand signalling regulation in human disorders, to develop network-based models underlining drug resistance and to suggest intervention sets. The resource is applicable not only to cancer-related studies, but also to study other pathologies. With the aim of revealing perturbations of molecular mechanisms in different human diseases, to predict drug sensitivity and to find optimal intervention schemes, one could combine the approaches represented in the example, into a workflow that contains the following steps: 1) Construction of comprehensive intra- and intercellular signalling network of a disease. 2) Integration of omics data and retrieval of network-based signatures, characterizing the disease. 3) On data availability, the resistance to treatment can be taken into account, and mechanisms associated with the resistance can be highlighted in a form of deregulated functional modules, pathways or key players. 4) Modelling mechanisms governing the disease and drug resistance and computing intervention gene sets to interfere with the disease and drug resistance. The omics data from each patient can be taken into account, to rank the intervention gene sets according to intrinsic vulnerabilities in each patient[12].

The suggested rationale has led to the creation of a collective research effort on different human disorders called the Disease Maps project (http://disease-maps.org).[19,20] The Disease Maps community aims at applying similar approaches as described in this chapter that will lead to identifying emerging disease hallmarks of various disorders, such as Parkinson's disease,[21] influenza[22] and others. This will help in studying disease comorbidities, to predict response to standard treatments and to suggest improved individual intervention schemes based on drug repositioning.[11]

## Useful Resources

https://acsn.curie.fr/ACSN2

https://acsn.curie.fr/ACSN2/downloads.html

https://www.vmh.life

http://www.home.ndexbio.org/index/

https://minerva.pages.uni.lu

http://disease-maps.org

http://www.garuda-alliance.org

## References

1. Hanahan, D. & Weinberg, R.A. Hallmarks of cancer: The next generation. *Cell* **144**, 646–674 (2011).
2. Chowdhury, S. & Sarkar, R.R. Comparison of human cell signalling pathway databases – Evolution, drawbacks and challenges. *Database* **2015**, 126 (2015).
3. Kuperstein, I. et al. Atlas of Cancer Signalling Network: A systems biology resource for integrative analysis of cancer data with Google Maps. *Oncogenesis* **4**, e160 (2015).
4. Kitano, H., Funahashi, A., Matsuoka, Y. & Oda, K. Using process diagrams for the graphical representation of biological networks. *Nat. Biotechnol.* **23**, 961–966 (2005).
5. Kondratova, M., Sompairac, N., Barillot, E., Zinovyev, A. & Kuperstein, I. Signalling maps in cancer research: Construction and data analysis. *Database* (2018).
6. Noronha, A. et al. OUP accepted manuscript. *Nucleic Acids Res.* **47**, D614–D624 (2018).
7. Kuperstein, I. et al. NaviCell: A web-based environment for navigation, curation and maintenance of large molecular interaction maps. *BMC Syst. Biol.* **7**, 100 (2013).
8. Bonnet, E. et al. NaviCell Web Service for network-based data visualization. *Nucleic Acids Res.* **43**, W560–W565 (2015). doi:10.1093/nar/gkv450.
9. Dorel, M., Viara, E., Barillot, E., Zinovyev, A. & Kuperstein, I. NaviCom: A web application to create interactive molecular network portraits using multi-level omics data. *Database* (2017).

10. Cerami, E. et al. The cBio cancer genomics portal: An open platform for exploring multidimensional cancer genomics data. *Cancer Discov.* **2**, 401–404 (2012).

11. Dorel, M., Barillot, E., Zinovyev, A. & Kuperstein, I. Network-based approaches for drug response prediction and targeted therapy development in cancer. *Biochem. Biophys. Res. Commun.* **464**, 386–391 (2015).

12. Monraz Gomez, L.C. et al. Application of Atlas of Cancer Signalling Network in preclinical studies. *Brief. Bioinform.* **20**, 701–716 (2018).

13. Costa, A. et al. Fibroblast heterogeneity and immunosuppressive environment in human breast cancer. *Cancer Cell* 33, 463.e10–479.e10 (2018).

14. Gawron, P. et al. MINERVA – A platform for visualization and curation of molecular interaction networks. *NPJ Syst. Biol. Appl.* **2**, 16020 (2016).

15. Pillich, R.T., Chen, J., Rynkov, V., Welker, D. & Pratt, D. NDEx: A community resource for sharing and publishing of biological networks. *Methods Mol. Biol.* **1558**, 271–301 (2017).

16. Ghosh, S., Matsuoka, Y., Asai, Y., Hsin, K.-Y. & Kitano, H. Toward an integrated software platform for systems pharmacology. *Biopharm. Drug Dispos.* **34**, 508–526 (2013).

17. Cerami, E.G. et al. Pathway commons, a web resource for biological pathway data. *Nucleic Acids Res.* **39**, D685–D690 (2011).

18. Kelder, T. et al. WikiPathways: Building research Communities on biological pathways. *Nucleic Acids Res.* **40**, D1301–D1307 (2012).

19. Mazein, A. et al. Systems medicine disease maps: Community-driven comprehensive representation of disease mechanisms. *NPJ Syst. Biol. Appl.* **4**, 21 (2018).

20. Ostaszewski, M. et al. Community-driven roadmap for integrated disease maps. *Brief. Bioinform.* **20**, 659–670 (2018).

21. Fujita, K.A. et al. Integrating pathways of Parkinson's Disease in a molecular interaction map. *Mol. Neurobiol.* **49**, 88–102 (2014).

22. Matsuoka, Y. et al. A comprehensive map of the influenza A virus replication cycle. *BMC Syst. Biol.* **7**, 97 (2013).

## CHAPTER 2

# Tumour Microenvironment Studies in Immuno- Oncology Research

## 2.1 NETWORK ANALYSIS OF THE IMMUNE LANDSCAPE OF CANCER

*Ajit J Nirmal and Tom C Freeman*

### 2.1.1 Summary

Targeting the immune component of tumours has been shown to hold great potential in the treatment of cancer. However, the composition of cells in the tumour microenvironment varies greatly both between different types of cancers and between tumours from a similar origin. To facilitate the identification and characterization of tumour-associated immune cells from transcriptomics data, a number of gene signatures have been reported, the majority of which have been defined through analysis of immune cells isolated from blood. However, blood cells do not necessarily reflect the differentiation or activation state of similar cells within tissues and consequently perform poorly as markers of immune cells in tissue. To address this issue, '*ImSig*', a set of immune cell gene signatures, were derived directly from tissue transcriptomics data using a network-based deconvolution approach. These enable the quantitative estimation of the immune content of tumour and non-tumour tissue samples, and our network-based analysis approach allows users to infer the context-dependent variations in the type or activation state of the immune cells within the tumour microenvironment. The *ImSig* algorithm is available as an R package ('imsig').

## 2.1.2 Introduction

Immunotherapy represents the most exciting advance in the treatment of cancer in the past decade. However, which patients respond to treatment is likely dependent on nature of tumour microenvironment, in particular the type and activation state of immune cells already present therein. Conventional methodologies such as immunohistochemistry are limited by the fact that histological analyses are restricted to the use of a small number of markers and flow cytometric analyses requires tissue disaggregation, which is often not practical. Hence, a number of computational approaches have been proposed to estimate the immune content of tissue biopsies from transcriptomic data.[1-5] All of these methodologies make use of immune signatures, i.e. lists of marker genes that together are indicative of the presence of a given immune cell population. A major shortfall of the available signatures is that they are generally derived from cells derived from the blood-derived cells. The expression profiles of the same immune cell from blood and tissues are significantly different[6] which compromises the predictive value of the signatures.[7] Additionally, between the signatures numerous markers are used interchangeably to define different immune subtypes and other 'markers' are clearly expressed by non-immune cell types.

To address these issues, we developed '*ImSig*', a set of immune gene signatures derived directly from tissue transcriptomics data using an unsupervised network-based deconvolution approach.[8] By analysing a broad range of human tissue transcriptomic data, a set of robustly co-expressed marker genes representing seven distinct immune cell types and three pathway systems were identified. The approach is based on the fact that genes that are involved in a biological process or whose expression is restricted to a particular cell type are frequently co-regulated, giving rise to expression 'modules' – groups of genes that in a given context exhibit highly similar profiles of expression. Such coexpression modules or clusters have been shown previously to be highly informative in many contexts (including cancer).[9-11]

## 2.1.3 Approach and Application Examples

*ImSig* contains 569 marker genes representative of seven immune populations [B cells (37 genes), plasma cells (14), monocytes (37), macrophages (78), neutrophils (47), NK cells (20), T cells (85)] and three biological processes [interferon response (66), translation (86), proliferation (99)]. This collection of gene lists can be used to estimate the relative abundance of individual immune cell types, as well as determine the nature of the tumour microenvironment from datasets of cancer transcriptomics data (or other tissue derived clinical datatsets). When a cell type is present in a sample set, the majority of marker genes for that cell should be observed to be coexpressed, i.e. they are highly correlated in their expression. In such circumstances the average expression of a gene signature can be used as a proxy for immune cell number, i.e. the greater the signal in a given sample, the more of that cell type is present. To aid these analyses we developed the R package 'imsig' which performs an additional pre-filtration (feature selection) step before computing the relative abundance of immune cells. This ensures that signature genes involved in other processes and therefore not coexpressed (in a given context), are excluded from the calculation of the average. An overview of the derivation and use of *ImSig* is shown in Figure 2.1.

To illustrate the potential to estimate the relative abundance of immune cells, a transcriptomics dataset derived from controls and trachoma patients was used.[12] The samples were taken from three patient subgroups; 20 controls with normal conjunctivas; 20 individuals with clinical signs of trachoma but that tested negative for the bacteria *C. trachomatis* (possibly who were in the resolution stage); and 20 individuals with symptoms and active infections were examined. This dataset was chosen due to the well-documented immune infiltration associated with this disease and the presence of all immune populations defined by *ImSig*. Our analysis showed there to be a significant increase in all immune populations associated with both patient groups relative to controls, particularly those patients with an active infection (Figure 2D, Nirmal et al.).[8] We went on to show that in addition to being able to explore changes in cell number, using network analysis, the nature of the immune microenvironment can also be studied. The *ImSig* signatures were used to identify other context-specific genes expressed by the immune cells present within. For example, in these data the T cell and macrophage signatures were correlated with each other and many others were immune-related. For example, the expression profile of genes such as *IFNG, LAG3, CD44, FOX03, FOXP3, CD80, IL20,*

Derivation of *ImSig*

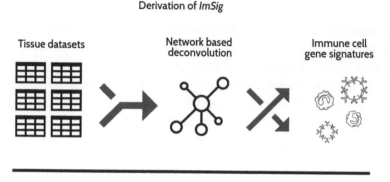

Implementation of *ImSig* in users dataset

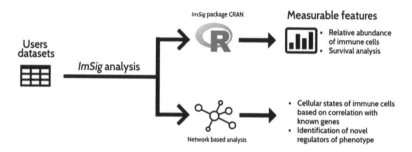

FIGURE 2.1 An overview of the derivation and use of *ImSig*. The upper panel describes the derivation of *ImSig*. Correlation-based network graphs were built from multiple tissue datasets and clustered using the Markov clustering algorithm (MCL). Modules of genes that represented various immune cells and cellular pathways were identified and filtered to define *ImSig*. The lower panel describes the implementation of *ImSig* in a users' dataset. One could use the R package ('imsig') for determining the relative abundance of immune cells among patient samples and in the event of availability of survival information, a survival analysis to determine the effect of immune infiltration in patient survival. On the other hand, the user could also perform a network-based analysis to identify cellular states of immune cells in the dataset of interest.

*STAT4, IL17A* etc. were correlated with T cell signature genes, indicating that the T cell population included Th17, Treg and Th1 subtypes. Similarly, genes associated with the macrophage signature contained many classical M1 markers. Taking this together it can be inferred that the samples exhibit the hallmarks of a classical inflammatory immune microenvironment, actively trying resolve the infection (Figure 3, Nirmal et al.).[8] In case of a cancer dataset, depending upon the samples under study, such context-specific genes could highlight pathways that are activated in drug resistance or vulnerabilities that could be targeted for converting non-responders into responders, or potentially used to identify molecular patterns that could be used to predict response.

To gain insights into the tumour microenvironment's role in patient prognosis, researchers often correlate immune infiltration with survival data. The most basic form of this analysis would involve grouping the patients into two groups with 'high' or 'low' immune cell infiltrate and subsequently performing a cox-proportional hazard ratio analysis. The hazard ratio score helps to inform a researcher as to the likely influence of immune cell infiltration with patient prognosis. In order to perform such analysis with *ImSig*, we have made it possible to calculate the average expression of the signature genes in each patient and then order the patients based on the resultant value. The patients can then be divided into two groups; usually based on the median value. *ImSig*'s R package has this feature in-built and so it is also possible to use the package for doing this analysis, quickly and easily. To illustrate this point, we obtained pre-normalized (level 3 data) transcriptomic data from 12 cancer-types from the TCGA (The Cancer Genome Atlas) database (Figure 2.4A, Nirmal et al.).[8] For each cancer type, patient samples were ordered based on the average expression of the individual *ImSig* signatures and split into two groups based on the median expression value of the signature genes. Whilst the survival analysis was not adjusted for potentially confounding variables (such as tumour stage, grade, age or treatment), the findings were largely consistent with the literature. In melanoma (SKCM), the known association between tumor infiltrating lymphocytes (TIL) and a good prognosis were observed[13] as well as in breast cancer (BRCA), where several studies have associated TILs with a good prognosis.[14] A negative association between TILs and prognosis was evident in low-grade glioma (LGG) and lung squamous cell carcinoma (LUSC), again in accordance with the previous literature.[15,16]

Lastly, *ImSig* can facilitate immune-subgrouping of patients based on the nature of the immune infiltrate, i.e. it is not only the number of immune cells present that potentially matters but their relative proportions. Understanding this heterogeneity may allow for the identification of those who are likely to respond to immunotherapy. For instance, we performed immune-subgrouping of samples of human skin cutaneous melanoma (SKCM) RNA-Seq data from the TCGA resource (Figure 2.1, Nirmal et al.).[8] Using *ImSig* genes as features we first generated a sample-to-sample correlation plot (r > 0.85). Following MCL clustering, the patients were grouped into five distinct clusters. These groupings were analyzed to study the expression patterns of immune cells within them. Nearly half the patients were in cluster-1, characterized by a low level of immune infiltrate. Hazard ratio (HR) analysis between these low immune (cluster-1) and high immune infiltrate (clusters-2 and -3) tumours revealed a significant difference in survival (HR: 0.38, p = 3E-9). Within the high immune subgroup, cluster-2 appeared to have a higher level of B cells and plasma cells in contrast to cluster-3 but overall survival (HR) was not significantly different between the two groups. Cluster-4 samples displayed higher levels of the interferon response genes and also showed improved survival compared to the low immune group. Whilst patients in clusters-2 and cluster-4 did not show a significant difference in hazard ratio compared to those in cluster-3, they could potentially show other features, such as differing responses to treatment. This analysis also highlights the level of immune heterogeneity in melanoma patients and its influence on survival.

### 2.1.4 Discussion and Perspectives

Researchers have long sought to define clinically meaningful tumour subtypes, based on pathological appearance, tissue of origin, marker expression and mutation profile, as means to predict their progression and response to therapy, but often with limited success. The primary reason for this is the degree of cellular heterogeneity between cancers, both in terms of the tumour cells themselves and the normal cells that both support and control their growth. Understanding and modelling this heterogeneity holds huge potential in the treatment of cancer. To this end, it is becoming increasingly apparent that one of the major influences on tumour growth, metastasis and drug responses is the host's immune system and how it responds to the tumour.

Whilst heterogeneity amongst the immune microenvironment of tumours undoubtedly exists, definition of robust sets of markers that can identify different immune cell types has proved challenging; there is little agreement between previous studies about the cell types associated with tumours and the genes that define them. Our studies have found that genes associated with a specific cell population or biological process form highly connected cliques of nodes, when large collections of data are subjected to network-based correlation analysis. By capitalizing this principle, it is possible to gain an insight into the complexity of tumour-immune interactions. We believe that *ImSig*, along with a network-based analysis framework, can help characterize the immune microenvironment of tumours and therefore explorations into the link between the immune system and a tumour's progression and response to therapy.

## Useful Resources

https://github.com/ajitjohnson/imsig

## References

1. Abbas, A.R. et al. Immune response in silico (IRIS): Immune-specific genes identified from a compendium of microarray expression data. *Genes Immun.* **6**, 319–331 (2005).
2. Abbas, A.R. et al. Deconvolution of blood microarray data identifies cellular activation patterns in systemic lupus erythematosus. *PLoS One* **4**, e6098 (2009).
3. Newman, A.M. et al. Robust enumeration of cell subsets from tissue expression profiles. *Nat. Methods* **12**, 453–457 (2015).
4. Becht, E. et al. Estimating the population abundance of tissue-infiltrating immune and stromal cell populations using gene expression. *Genome Biol.* **17**, 218 (2016).
5. Bindea, G. et al. Spatiotemporal dynamics of intratumoral immune cells reveal the immune landscape in human cancer. *Immunity* **39**, 782–795 (2013).
6. Schelker, M. et al. Estimation of immune cell content in tumor tissue using single-cell RNA-seq data. *Nat. Commun.* **8**, 2032 (2017).
7. Pollara, G. et al. Validation of immune cell modules in multicellular transcriptomic data. *PLoS One* **12**, e0169271 (2017).
8. Nirmal, A.J. et al. Immune cell gene signatures for profiling the microenvironment of solid tumors. *Cancer Immunol. Res.* **6**, 1388–1400 (2018).
9. Doig, T.N. et al. Coexpression analysis of large cancer datasets provides insight into the cellular phenotypes of the tumor microenvironment. *BMC Genom.* **14**, 469 (2013).
10. Freeman, T.C. et al. A gene expression atlas of the domestic pig. *BMC Biol.* **10**, 90 (2012).
11. Forrest, A.R.R. et al. A promoter-level mammalian expression atlas. *Nature* **507**, 462–470 (2014).
12. Natividad, A. et al. Human conjunctival transcriptome analysis reveals the prominence of innate defense in *Chlamydia trachomatis* infection. *Infect. Immun.* **78**, 4895–4911 (2010).

13. Ladanyi, A. Prognostic and predictive significance of immune cells infiltrating cutaneous melanoma. *Pigment Cell Melanoma Res.* **28**, 490–500 (2015).
14. West, N.R. et al. Tumor-infiltrating FOXP3(+) lymphocytes are associated with cytotoxic immune responses and good clinical outcome in oestrogen receptor-negative breast cancer. *Br. J. Cancer* **108**, 155–162 (2013).
15. Yao, Y. et al. B7-H4(B7x)-mediated cross-talk between glioma-initiating cells and macrophages via the IL6/JAK/STAT3 pathway lead to poor prognosis in glioma patients. *Clin. Cancer Res.* **22**, 2778–2790 (2016).
16. Hiraoka, K. et al. Inhibition of bone and muscle metastases of lung cancer cells by a decrease in the number of monocytes/macrophages. *Cancer Sci.* **99**, 1595–1602 (2008).

## 2.2 INTEGRATIVE CANCER IMMUNOLOGY AND NOVEL CONCEPTS OF CANCER EVOLUTION

*Jérôme Galon, Mihaela Angelova, Bernhard Mlecnik, Gabriela Bindea and Daniela Bruni*

### 2.2.1 Summary

All proposed models describing tumour development and progression ultimately rely on the centrality of tumour cell features in guiding and shaping tumour evolution, up to the metastatic spreading. The central role played by the immune system in the effector phases leading to tumour eradication has been recognized nowadays; nonetheless, its contribution to the steps defining tumour evolution has been less appreciated. We recently assessed the impact of the immune system in metastatic evolution in humans. To do so, we relied on multi-omics approaches on multiple lung, liver or peritoneal synchronous and metachronous metastases, as well as primary tumours from colorectal cancer patients. We revealed a highly heterogeneous metastatic landscape, in which non-recurrent clones are immunoedited, whilst progressing clones are immune-privileged, thereby first demonstrating how immunoediting impacts human cancers progression. Immunoediting and Immunoscore were the two best predictors of a favourable clinical outcome, and, together with metastatic size, could independently predict recurrence. We proposed a parallel-multiverse immune selection tumour evolution model reflecting the influence of the immune system on tumour heterogeneity and clonal evolution via the immunoediting process. This chapter describes the methodology and the analytical pipelines employed to achieve this paradigm shift, as well as the obtained results.

### 2.2.2 Introduction

Recent scientific advances in the field of oncology and the success of anti-cancer therapies targeting the patient immune system provided us with an unprecedented level of understanding of tumour dynamics. The concept that immune reactions within the tumour sites influence clinical outcome[1-3] encouraged the adoption of a renewed vision of malignant cells, no longer seen as individual entities but as part of a dynamic microenvironment. Whilst the role of the immune system in the effector phases eventually leading to tumour elimination have been widely recognized,[4] its contribution to the steps defining tumour evolution has been less appreciated. Cancer cells are well known for carrying genetic mutations. In 1990, Fearon and Vogelstein[5] presented a multistep model for the development of colorectal cancers (CRC) based on the temporally sequential accumulation of specific genetic mutations conferring a selective advantage to the correspondent clone. This view was challenged by Sottoriva et al. in 2015,[6] who proposed the so-called 'big-bang model'. According to this model, tumours derive from the expansion of an initial single clone, implying the simultaneous presence of the initial mutations in all tumour cells. Perhaps the most widely accepted model nowadays is the 'branched evolution model' proposed by Gerlinger et al. in 2012,[7] according to which multiple clones expressing distinct mutations and deriving from a common mutated ancestor co-evolved in parallel, and co-exist. Of note, at the core of these and additionally proposed theories[8] is the intrinsic concept of the centrality of tumour cell features in driving and directing tumour evolution. A recent work by Angelova et al. sought to apply a Darwinian approach to investigate the role of the immune system in this process. The adoption of such broader angle enabled the proposal of a novel model of cancer evolution and metastatic progression incorporating the role of the immune system, defined parallel-multiverse immune selection model, thereby shifting from a 'tumour-cell centric' to an 'immune-centric' vision of human cancer.

## 2.2.3 Approach and Application Example

Using systems tumour-immunology, we previously showed the spatio-temporal dynamics of intratumoural immune cells, first revealing the *immune landscape* in human cancer.[9] Using immune gene expression from purified immune sub-population, we defined the *Immunome*, and applied it to human tumours, allowing to decipher immune cell infiltration from gene expression profiles from a complex mixture of cells. The results showed that the adaptive immune response within tumours was maximal at the earliest stage of carcinoma (T1 stage). Thanks to our network visualization software, ClueGO[10] and CluePedia,[11] we deciphered the mechanisms associated with increased density of immune cell subsets within tumours.[12] We previously showed that tumour recurrence and overall survival times were mostly dependent on the presence of cytotoxic and memory T cells within specific regions of primary tumours.[1] The importance of the patient intratumour natural adaptive immune reaction for the survival of patients revealed that immune parameters are beyond tumour progression and invasion (TNM classification). We hence developed and validated the Immunoscore, a consensus IHC- and digital pathology-based scoring system based on the quantification of CD3+ and CD8+ cells in two distinct regions of the tumour, the centre and the invasive margin.[1,13] We demonstrated that the tumour microenvironment and Immunoscore® within primary tumours are critical determinants of dissemination to distant metastasis.[14]

The work by Angelova and colleagues was based on the study of two rare cases of metastatic CRC with an exceptional 11-year clinical follow-up. These patients underwent surgical resection of their primary tumours, as well as of liver, lung and peritoneal metastases over time. Deep biological and bioinformatics analyses were performed on a total of 36 FFPE samples, thereby enabling building of a model for tumour evolution that was then validated on further 126 samples. Both genomics and immunomics approaches were applied.

Whole-exome sequencing with a 180X average sequencing coverage was performed on matched blood samples, resected primary tumours and consecutive metastases. The resulting data were subjected to multiple types of analyses. First, the total number of coding and non-coding somatic mutations was obtained, also in combination with the targeted sequencing of 50 frequently mutated cancer genes. This analysis revealed the tumour mutational spectrum associated with each sample and showed a high inter-metastatic genetic divergence. Indeed, mostly unique coding mutations were found in each metastasis, and a very low number of shared mutations amongst individual patients' samples. Second, the Parsimony Ratchet method was applied to build a phylogenetic tree based on the concordance of non-silent point mutations among the primary tumours and metastases. Interestingly, metastases in the same anatomic locations were phylogenetically closer and shared more mutations compared to distant metastases. Third, the similarity of the coding mutations profiles was exploited to infer the origin, i.e. the parent–child relationship for each metastasis, as defined by the highest Jaccard similarity coefficient. The clonal distribution within each sample was reconstructed. The integration of the sequencing coverage with the inferred parental copy numbers and estimated tumour purities yielded the cellular prevalence of a specific somatic mutation[15]; tumour clones were then inferred based on a cross-sample Bayesian clustering approach.[16] For a subclonal hierarchical classification, the genetic algorithm SCHISM v1.1.2 was applied to the cellular and mutation frequencies. Overall, these analyses allowed tracing of the route followed by individual tumour clones, as well as their aggressiveness. Some clones were found to spread inter-metastatically, even to distant organs. These results were collectively displayed in a metastatic evolvogram displaying the samples by detection time. Apart from showing a high intrametastatic clonal heterogeneity, the evolvogram shows how some clones persisted, and disseminated, whilst others disappeared. Further analyses revealed that the progressing clones were immune-privileged (i.e. not subjected to immune attack), whilst non-recurrent clones (<4 years) were immunoedited, as indicted by their low immunoediting score (the ratio of expected to observed immunogenic mutations per non-silent mutation). Of note, the immunoediting score was calculated via a newly developed method,[15] based on a previously reported approach.[17] Fourth, the realignment of the exome sequencing reads on the unspliced genomic regions of several TCR and Ig chains, which were converted into the correspondent T cell and B cell clonotypes via the MiXCR tool. Hence, for each patient, the T and B cell populations were identified and quantified.

To comprehensively reveal the contribution of the immune system in this process, the immune landscape of each lesion was evaluated qualitatively, quantitatively and spatially by the means of single-plex IHC staining, digital quantification and multispectral imaging for several immune markers of major lymphocytic lineages, including CD3[+], CD8[+], FoxP3[+] and CD20[+]; Ki67[+] (a cellular marker for proliferation) and cytokeratin (for epithelial cells) were also included in the multiplex analysis. This approach also enabled calculation of the Immunoscore associated to each metastasis/tumour. Plotting the Immunoscore values (High *vs* Low, henceforth displayed as Hi/Lo) against the corresponding immunoediting score (Yes/No) for each metastatic sample generated four categories: HiYes, HiNo, LoYes and LoNo. No metastasis fell within the LoYes group, reinforcing the hypothesis that T cells are required for immunoediting to occur. Accordingly, the majority of metastases fell in the LoNo category. 54% of the metastases with high Immunoscore were immunoedited. Overall, these results provide the first direct evidence of immunoediting in humans at the metastatic stage, and indicate that T cell infiltration is necessary but not sufficient for immunoediting to occur. When analysing recurrence amongst these four categories, the LoNo were found to produce at least one child metastasis (64% of cases), whilst that was the case for only 8% of the high Immunoscore metastases.

Single-plex and multiplex IHC, as well as immune gene expression analysis by Nanostring technology, were instrumental in revealing the correlation between T cell infiltration/Immunoscore and the TCR-alpha diversity, as determined by whole exome sequencing. Multispectral imaging analysis allowed for a pairwise comparison of all cell phenotypes in terms of mutual neighbour distances and nearest-neighbour distance distribution functions ($G_{km}$). Such spatial analysis revealed important determinants correlated with metastatic spreading. Indeed, Cox multivariate analysis revealed four parameters associated with metastatic dissemination: 1) the Immunoscore, 2) the immunoediting, 3) the distance between CD3+ T-cells and proliferating (Ki67+) tumour (CK+) cells and 4) the size of the parent metastasis (Figure 2.2). These four parameters were found to be all independent predictive factors of metastatic recurrence. Specifically, large metastases with low Immunoscore and no immunoediting displayed the highest probability of recurrence, which was found particularly pronounced in the first two years after detection. Based on these covariates, a predictive time-to-event model of metastatic cancer evolution was built, and for each metastasis the cumulative recurrence probability over time was calculated. This model rightly predicted a high risk of recurrence for one patient, which indeed relapsed with several unresectable metastatic lesions. A low risk of recurrence was found for another patient, which remains to date in complete remission. The model was validated independently on a case study on ovarian cancer, and on 132 primary CRCs to predict the first event of metastatic dissemination. In all cases, the model could accurately stratify tumours

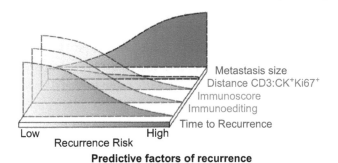

**Predictive factors of recurrence**

FIGURE 2.2    Predictive factors of metastatic recurrence. Schematic representation of the influence of metastatic size, distance CD3+: CK+ Ki67+ cells (T cells and proliferating tumor cells, respectively), Immunoscore and immunoediting on the recurrence risk associated to a given metastasis. Big metastases with low associated Immunoscore and immunoediting are more likely to recur than small metastases with high Immunoscore and immunoediting. (Adapted from Angelova, M. et al. Evolution of metastases in space and time under immune selection. *Cell* **175**, 751.e716–765.e716 [2018].)

into low, intermediate and high risk, with confirmed significantly different recurrence rates.

## 2.2.4 Discussion and Perspectives

The described integrative approach represents the merging of distinct fields, encompassing Genetics, Immunology, Oncology and Bioinformatics. By tracking metastases evolution in space and time under immune selection, this study revealed how the history of tumour clones is tightly shaped by the intra-metastatic immune contexture. It also proves for the first time in a human metastatic setting the power of immunoediting in determining clone recurrence, together with metastatic size, Immunoscore and the proximity of T cells to proliferating tumour cells. Hence, the proposed parallel immune selection model represents the first comprehensive predictive tool base on a global tumour *plus* immune dynamics assessment. As a matter of fact, the underlying concept is not so novel after all, as it results from the application of an exquisitely Darwinian theory: The better a tumour clone adapts to its microenvironment, the higher the survival advantage and persistence in subsequent metastases.

It is more and more evident that the deciphering of the complex interplay between the tumour and immune system requires the contribution of different technical and bioinformatics approaches. Only comprehensive analysis holds the potential to yield an accurate scheme of the driving forces behind a complex mechanism such as that underpinning metastatic dissemination. It is likely that the addition of further parameters, for example coming from a higher multiplex ability, might pinpoint additional key determinants of metastatic spreading, thus improving the precision of the obtained model. This would also represent a step forward towards the concept of precision medicine, although a methodological simplification and concomitant reduction of the associated experimental costs are needed to facilitate clinical translation of these findings. In addition, metastases are not always resectable, and resected. In this sense, the recent report of the high accuracy of the Immunoscore also when performed on biopsies[18] represents an encouraging step in this direction.

## Useful Resources

http://www.ici.upmc.fr/

https://apps.cytoscape.org/apps/cluego

https://apps.cytoscape.org/apps/cluepedia

http://www.immunoscore.org/

## References

1. Galon, J. et al. Type, density, and location of immune cells within human colorectal tumors predict clinical outcome. *Science* **313**, 1960–1964 (2006).
2. Galon, J., Fridman, W.H. & Pages, F. The adaptive immunologic microenvironment in colorectal cancer: A novel perspective. *Cancer Res.* **67**, 1883–1886 (2007).
3. Pages, F. et al. Effector memory T cells, early metastasis, and survival in colorectal cancer. *N Engl J Med.* **353**, 2654–2666 (2005).
4. Chen, D.S. & Mellman, I. Oncology meets immunology: The cancer-immunity cycle. *Immunity* **39**, 1–10 (2013).
5. Fearon, E.R. & Vogelstein, B. A genetic model for colorectal tumorigenesis. *Cell* **61**, 759–767 (1990).
6. Sottoriva, A. et al. A Big Bang model of human colorectal tumor growth. *Nat Genet.* **47**, 209–216 (2015).
7. Gerlinger, M. et al. Intratumor heterogeneity and branched evolution revealed by multiregion sequencing. *N Engl J Med.* **366**, 883–892 (2012).
8. Williams, M.J., Werner, B., Barnes, C.P., Graham, T.A. & Sottoriva, A. Identification of neutral tumor evolution across cancer types. *Nat Genet.* **48**, 238–244 (2016).
9. Bindea, G. et al. Spatiotemporal dynamics of intratumoral immune cells reveal the immune landscape in human cancer. *Immunity* **39**, 782–795 (2013).
10. Bindea, G. et al. ClueGO: A Cytoscape plug-in to decipher functionally grouped gene ontology and pathway annotation networks. *Bioinformatics* **25**, 1091–1093 (2009).
11. Bindea, G., Galon, J. & Mlecnik, B. CluePedia Cytoscape plugin: Pathway insights using integrated experimental and in silico data. *Bioinformatics* **29**, 661–663 (2013).
12. Mlecnik, B. et al. Biomolecular network reconstruction identifies T-cell homing factors associated with survival in colorectal cancer. *Gastroenterology* **138**, 1429–1440 (2010).
13. Pages, F. et al. International validation of the consensus Immunoscore for the classification of colon cancer: A prognostic and accuracy study. *Lancet* **391**, 2128–2139 (2018).
14. Mlecnik, B. et al. The tumor microenvironment and Immunoscore are critical determinants of dissemination to distant metastasis. *Sci Transl Med.* **8**, 327ra326 (2016).
15. Angelova, M. et al. Evolution of metastases in space and time under immune selection. *Cell* **175**, 751.e716–765.e716 (2018).
16. Roth, A. et al. PyClone: Statistical inference of clonal population structure in cancer. *Nat Methods* **11**, 396–398 (2014).

17. Rooney, M.S., Shukla, S.A., Wu, C.J., Getz, G. & Hacohen, N. Molecular and genetic properties of tumors associated with local immune cytolytic activity. *Cell* **160**, 48–61 (2015).
18. Van den Eynde, M. et al. The link between the multiverse of immune micro-environments in metastases and the survival of colorectal cancer patients. *Cancer Cell* **34**, 1012–1026 (2018).

## 2.3 SYSTEMS BIOLOGY APPROACH TO STUDY HETEROGENEITY AND CELL COMMUNICATION NETWORKS IN THE TUMOUR MICROENVIRONMENT

*Floriane Noël and Vassili Soumelis*

### 2.3.1 Summary

Cell–cell communication is at the basis of the higher order organization observed in tissues, organs and organisms. Understanding cell–cell communication, and its underlying mechanisms that drive the development of cancer is essential. The tumour microenvironment is composed of a great cellular diversity, such as endothelial, stromal or immune cells that can influence tumour progression as well as its response to treatment.

This chapter is focused on deciphering the heterogeneity of the tumour microenvironment and its impact on cell–cell communications. To do so, we use systems biology approaches and communication network reconstruction to better understand the interplay between immune cell subsets and tumour cells.

### 2.3.2 Introduction

*Concepts of Communication*

Communication involves an exchange of information between two entities or components of the system. In biological systems, communication is largely, but not exclusively, mediated by chemical signals that are sent by a cell and received and processed by another cell. This results in triggering a specific response, which can be very variable in its quality and intensity depending on the nature and dose of the stimulus.

In living systems, communication is essential to critical processes such as development, growth, survival, maintenance and defence of the system (from individual cells to whole organisms). These complex functions require a tight and very well-regulated communication between the diversity of cell types. Intercellular communication can be viewed and formalized as networks in which nodes are cells and edges represent communication signals. The originality of such networks is their multilevel complexity, since cells can belong to various subsets and harbour multiple states, and they can communicate with each other by exchanging a multiplicity of information signals.

*Steady State versus Inflammation*

In living systems theory, the steady state is defined as requiring the minimal exchange of information between the system and its environment.[1] In the presence of stress, the system needs to adapt, which is accompanied by an increase in information exchange between systems components, and between the system and its environment.

Inflammation can be seen as the reaction of the tissue or organ to any type of stress (physical, chemical, infectious). In a physiological setting, the inflammatory reaction takes an acute form and is accompanied by wound healing and tissue repair mechanisms, together with an appropriate immune response, collectively leading to restoration of the steady-state. On the contrary, pathological inflammation can lead to a chronic disease due to an excess or insufficiency of the tissue repair and immune mechanisms.

### Cancer as a Dysregulation of Cell–Cell Communication

Cancer is a form of pathological inflammation in which immune and tissue repair mechanisms fail in: (1) eliminating malignant tumour cells and (2) restoring the normal tissue architecture and function. The physiopathology of cancer is very complex and several theories and views have been proposed, each one putting an emphasis on a different aspect of the disease. For example, cancer includes acquired regulations in key physiological processes inside tumour cells, a disruption in normal tissue architecture and an escape from the immune system. Here, we want to propose a view that cancer is also a disease of inappropriate communication between the different cellular players accounting for the tumour microenvironment: malignant tumour cells, normal epithelial cells, fibroblasts, innate and adaptive immune cells. Deciphering the complexity of cell communication networks in cancer is a great challenge and should take into account the multiplicity of cell types/states, and of communication signals characterizing each type of tumour inflammation. We will present some original approaches that have been developed to analyze cellular heterogeneity and intercellular communication networks by combining experimental and computational methods.

### 2.3.3 Approach and Application Example

Given the complexity of cell communication networks, involving a multiplicity of cell types and cell states as nodes, and the multiplicity of communication molecules as edges (Figure 2.3), our approach is to initially dissociate cellular and molecular heterogeneity, followed by methods to connect those various components into integrated cell communication networks.

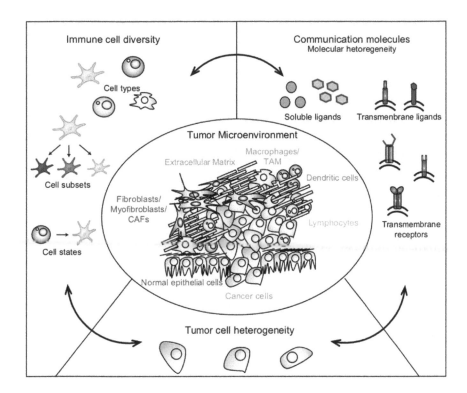

FIGURE 2.3 Cell–cell communication in the tumour microenvironment. Schematic representation of the tumour microenvironment composed of various cell types. Important concepts of cell–cell communication in the tumour microenvironment are the immune cell diversity (upper left), the communication molecules and their molecular heterogeneity (upper right) and the tumour cell heterogeneity (bottom). These three parts are interconnected and are essential to reconstruct communication networks in the tumour microenvironment.

## Cell Heterogeneity

Metazoans are composed of different cell types from pluripotent stem cells to tissue-specific cells, harbouring specialized functions. In cancer, heterogeneity exists in all different cell types composing the tumour.

*Tumour cell heterogeneity* has been described as deriving from cancer stem cells (CSC)[2] as well as clonal evolution and selection.[3] Intra-tumoural heterogeneity is a complex interplay between CSCs genetic and epigenetic mutations and clonal evolution promoting development and evolution of cancer to metastasis.[4] Regarding cell–cell communication, studies have highlighted the link between specific mutations in cancer cells and alteration of critical signalling pathways driving mechanisms of tumour cell survival, growth and metastasis.[5]

*Immune cell heterogeneity:* The tumour microenvironment is composed of a large diversity of immune cell types and subsets (Figure 2.3). Cellular crosstalk has been already identified as being associated to pro- or anti-tumour immunity, already pointing at intercellular communication as a critical aspect of tumour evolution.[6,7] Such studies have focused on crosstalk between predefined cell types. We have set up an original program to analyze the global cellularity and heterogeneity in human primary tumours (TMEGA: Tumour MicroEnvironment Global Analysis) using a combination of flow cytometry and medium scale proteomics measurements. This led to defining the cellular composition of the breast cancer microenvironment on over 400 individual tumour samples by quantifying various immune cell types and subsets. In parallel, we and others are implementing complementary approaches in order to get a more in-depth and qualitative evaluation of immune cell states, which can be linked to functional aspects including intercellular communication. We have completed a transcriptomics analysis of human dendritic cell subsets in primary luminal and triple negative breast cancer,[8] which revealed various transcriptional states and signatures between subsets, as well as different transcriptional programs of dendritic cell adaptation to the tumour microenvironment.[8] Recent studies using the scRNAseq technique identified a diversity of dendritic cell (DC) subsets and states in breast and head and neck cancers.[9,10] Such studies enriched our view of DC diversity and plasticity leading to functional specialization in inflammatory conditions.[11]

## Molecular Heterogeneity

Molecular heterogeneity is the basis of information exchange between different cells through ligand-receptor pairs. Communication molecules can be membrane-bound, implying cell–cell contact, or soluble proteins or lipid mediators (Figure 2.3). Each family of communication molecules, for example cytokines, is itself very diverse and involved in regulating a large number of physiological and pathological processes. Very few studies have focused on a systematic description of communication molecules in cancer. In our TMEGA study, we have exploited tumour conditioned supernatants in order to measure 50 soluble analytes selected for their implication in communication within the tumour microenvironment. We have identified molecules that are differentially expressed between the tumour and non-involved juxtatumour tissue, again confirming the importance of dysregulated communication as a main feature of tumour inflammation.

*Connection Ligand/Receptors*

Once the heterogeneity of cell states and communication molecules has been estimated using various data types, the challenge is to infer effective cell–cell communication. One approach is to exploit knowledge databases to associate communication 'ligands' to their cognate receptors. Several such resources have been published,[12-14] and we developed our own manually curated database (ICELLNET platform) focusing on immune molecules (F. Noel et al., unpublished). ICELLNET is a platform for cell communication network reconstruction from cell transcriptomics profiles. It can predict intercellular communication channels by Ligand-Receptor coupling. Similar methods have been applied to scRNAseq data,[15,16] and could identify regulatory interactions between different cell types helping to better characterize cell–cell communication in specific environments.

The next step would be to integrate various data types in addition to transcriptomics, in particular FACS and proteomics data, as assessed in our TMEGA program.

### 2.3.4 Discussion and Perspectives

Cell communication networks are of tremendous complexity, and a diversity of complementary approaches are needed to infer physio-pathologically relevant networks. Bioinformatics and computational methods are required to handle the analyses, exploit large-scale datasets and integrate cell and molecular data into comprehensive network models. Single cell technology should add valuable information to current knowledge.

Spatial and temporal heterogeneity in cell–cell communication is still insufficiently studied, and should be a focus in future studies, despite intrinsic difficulties and methodological bottlenecks. Intercellular communication networks could guide therapeutic targeting of the most relevant molecular and cellular components characterizing tumour inflammation.

### References

1. Miller, J.G. Living systems. *Curr. Mod. Biol.* **4**, 55–256 (1972).
2. Meacham, C.E. & Morrison, S.J. Tumor heterogeneity and cancer cell plasticity. *Nature* **501**, 328–337 (2013).

3. Greaves, M. & Maley, C.C. Clonal evolution in cancer. *Nature* **481**, 306–313 (2012).

4. Torres, L. et al. Intratumor genomic heterogeneity in breast cancer with clonal divergence between primary carcinomas and lymph node metastases. *Breast Cancer Res. Treat.* **102**, 143–155 (2007).

5. Martin, G.S. Cell signalling and cancer. *Cancer Cell* **4**, 167–174 (2003).

6. Chen, D.S. & Mellman, I. Oncology meets immunology: The cancer-immunity cycle. *Immunity* **39**, 1–10 (2013).

7. DeNardo, D.G., Andreu, P. & Coussens, L.M. Interactions between lymphocytes and myeloid cells regulate pro-versus anti-tumor immunity. *Cancer Metastasis Rev.* **29**, 309–316 (2010).

8. Michea, P. et al. Adjustment of dendritic cells to the breast-cancer microenvironment is subset specific. *Nat. Immunol.* **19**, 885–897 (2018).

9. Azizi, E. et al. Single-cell map of diverse immune phenotypes in the breast tumor microenvironment. *Cell* **174**, 1293.e36–1308.e36 (2018).

10. Abolhalaj, M. et al. Profiling dendritic cell subsets in head and neck squamous cell tonsillar cancer and benign tonsils. *Sci. Rep.* **8**, 8030 (2018).

11. Soumelis, V., Pattarini, L., Michea, P. & Cappuccio, A. Systems approaches to unravel innate immune cell diversity, environmental plasticity and functional specialization. *Curr. Opin. Immunol.* **32**, 42–47 (2015).

12. Lizio, M. et al. Update of the FANTOM web resource: High resolution transcriptome of diverse cell types in mammals. *Nucleic Acids Res.* **45**, D737–D743 (2017).

13. Ramilowski, J.A. et al. A draft network of ligand–receptor-mediated multicellular signalling in human. *Nat. Commun.* **6**, 7866 (2015).

14. Kirouac, D.C. et al. Dynamic interaction networks in a hierarchically organized tissue. *Mol. Syst. Biol.* **6**, 417 (2010).

15. Vento-Tormo, R. et al. Single-cell reconstruction of the early maternal–fetal interface in humans. *Nature* **563**, 347–353 (2018).

16. Cohen, M. et al. Lung single-cell signalling interaction map reveals basophil role in macrophage imprinting. *Cell* **175**, 1031.e18–1044.e18 (2018).

## CHAPTER 3

# Multi-Level Data Analysis in Cancer

## *Tools and Approaches*

## 3.1 THE CYTOSCAPE PLATFORM FOR NETWORK ANALYSIS AND VISUALIZATION

*Benno Schwikowski*

### 3.1.1 Summary

Cytoscape is an open-source software platform that supports the visualization and analysis of molecular profiling data in the context of functional interaction networks. It is developed by several research groups that are actively involved in the development of technologies around the generation and integrative analysis of molecular profiling data in the context of biological and biomedical research. Here, we outline the rationale behind the use of functional interaction networks, introduce the Cytoscape platform, and present an example in which data analysis and visualization using Cytoscape has led to a discovery of previously unknown disease biology.

### 3.1.2 Introduction

Biology is organized hierarchically, from lower levels that comprise DNA, RNA and proteins, over cells, to multicellular organisms and ecosystems.

Changes in lower levels lead to changes in higher levels, and evolutionary forces that act on higher levels shape, over longer time spans, lower levels. Many research questions concerning complex diseases, such as cancer, thus require data across different genes, and across several biological levels. Functional interaction networks offer a meaningful context for molecular measurements, and thus interpretability of the strongly increasing number of molecular profiling datasets that are becoming accessible, in particular for cancer (e.g. Tomczak et al.[1]).

### Molecular Pathways and Functional Interaction Networks

Cancer is characterized by a distinct set of biological capabilities, for example resistance to cell death, or the ability to sustain proliferative signalling.[2] Most of these capabilities can be understood at the level of cellular and molecular physiology, and their acquisition is thought to arise through genetic instability.

To study how genetic instability leads to physiological change, it is natural to consider the intermediate level of molecular pathways that implement subcellular functions using systems of interacting molecules: DNA, RNA, proteins and metabolites. Measuring, or even defining 'pathway activity' at a large scale is difficult. Based on global profiling technologies, such as next-generation sequencing, one can, for example, try to derive measures of pathway activity from integrated measures of abundance of the underlying molecules (e.g. Martignetti et al.[3]). Databases, such as KEGG[4] and Reactome,[5] provide access to the molecules that constitute pathways, and their molecular interactions. However, our knowledge of pathways is still highly incomplete, and systematic approaches to determine unknown pathways do not exist. Therefore, global and exhaustive characterizations of biology at the pathway level, as it is now possible for DNA, RNA and proteins, are currently out of reach.

An alternative mode of analysis at the level between molecules and cellular physiology is based on *functional interaction networks*. Two molecules are said to interact functionally, whenever they potentially occur together in some functional context. Globally one can assemble these interactions into functional interaction networks that contain molecular pathways as connected subnetworks. Thus, to explore the possible bases of change in cellular physiology, one can look for aggregate changes in data about molecular subnetworks as reflections of change in molecular pathways.

Functional interaction networks can be based on various types of evidence. Genes and metabolites involved in sequential metabolic interactions provide one type of functional interaction. The resulting metabolic networks are particularly useful for integrated analysis, as they are relatively complete, and quantitative aspects of their function are relatively well understood. Popular databases of functional interactions[6,7] gather evidence for metabolic interactions, and a variety of other functional interactions, for example, physical protein–protein interactions from high-throughput, computational predictions and evidence from automated text mining.

As of today, most types of functional interaction networks have to be considered highly incomplete. Still, placing molecular profiling data in the context of functional networks does offer practical opportunities to identify and study changes in molecular pathways as changes in – potentially fragmentary – subnetworks of functional interaction networks.

### 3.1.3 Approach and Application Example
#### A Brief Tour of Cytoscape

The Cytoscape software platform[8,9] offers a wide range of functionality around the visualization and analysis of functional and other networks and network-associated data, with no need for programming. Cytoscape has been developed as an open-source, community-driven software platform since its first release in 2002. Initiated at the Institute for Systems Biology in Seattle by the need to explore some of the first large-scale protein interaction datasets, Cytoscape is now a software platform whose core is maintained and developed by a handful of different systems biology research teams and extends to a worldwide user community. Cytoscape is used across a wide range of applications, but its focus remains on biomedical research, and specifically, on molecular interaction networks.

Cytoscape consists of two parts: The core software, which provides a graphical user interface and a basic set of features for analysis and visualization, and apps; software with specific additional functions that can be added and removed as required. Cytoscape can hold one or several *Cytoscape networks*. Each Cytoscape network consists, on the one hand, of a *graph* that consists of a set of *nodes* and *edges* that connect pairs of nodes. Edges can be undirected or directed towards one of the nodes. On the other hand, edges and nodes have *attributes* in user-accessible node and edge tables. Attributes are numerical (e.g. values for gene expression), categorical (e.g. from a limited number of strings that code different kinds of protein–protein interaction) or strings (e.g. URLs pointing to publications with in-depth information about genes and their interactions). Attributes can be interpreted by user-configurable apps, thus allowing a wide and flexible range of use cases.

Cytoscape can import data directly from local files and from the internet. Cytoscape and its apps provide access to a wide range of internet databases with networks; experimental data. File formats include human-readable and human-editable formats for the network and associated data. Cytoscape can store and retrieve the current state of a session in *session files* that can be shared over the internet.

The Cytoscape user interface provides a number of panels for interacting with networks, node, edge and network tables, and the set of all networks in memory. The main network panel graphically represents the nodes and edges of the currently selected network. The graph layout can be modified directly in the network panel or optimized by a large selection of graph layout algorithms. Data in node and edge tables can be represented using various visual node and edge attributes. How each type of data is mapped to visual attributes can be configured and controlled by the user by means of *styles*; combinations of parameterized visual properties whose parameters correspond to node- or edge-specific data in the node and edge tables. Styles can be saved independently of data, and, once defined, reused in subsequent Cytoscape sessions. More than 20 accessible node properties include fill colour, shape, label, label colour and shape. The way in which node and edge attributes are mapped to visual properties, i.e. colour gradients for underlying numerical data, can be configured by the user. These features allow for flexible, largely automated and uniquely customized and information-rich interactive visualization of larger networks, beyond biological applications.

To provide further flexibility and extensibility by the Cytoscape user community for the rapidly growing use cases, Cytoscape uses a modular software architecture that allows most Cytoscape functionality to be extended by *apps*; optional software components. While apps can be installed, upgraded and uninstalled from within Cytoscape, most apps are accessed using the web-based Cytoscape App Store (Lotia et al.[10]; http://apps.cytoscape.org). The App Store offers browsing of apps by category, such as data import/export, visualization, data analysis and automation. For each of the over 300 apps available today, the App Store provides an author-curated web page with a searchable description as well as release and download information. Apps can be installed directly into a running Cytoscape session with just a mouse click. App pages also contain links to other app-specific information, such as journal articles and online discussions among users.

Cytoscape beginners can find links to manuals, tutorials, journal articles and presentations on the main Cytoscape website (http://cytoscape.org).

App developers starting out can find introductions to the Cytoscape architecture and APIs. Several searchable mailing lists provide exchange platforms for users and developers, and Cytoscape core developers participate to help with the hardest technical questions. Regular Cytoscape workshops and symposia bring user and developer communities together.

We now illustrate the analysis and visualization of subnetworks by a discovery of a new cell-physiological effect in a developmental disorder.[11] Briefly, the hallmarks of *cavernous cerebral malformation* (CCM), can be recapitulated in murine models by inactivating homologs of the genes CCM1, CCM2 and CCM3, which carry the causative mutations for CCM.[12,13] Transcriptomic profiles had been obtained from the relevant venous tissue in which causative genes were invalidated, and from controls.

Figure 3.1 shows a functional gene subnetwork, visualized in Cytoscape, and, on the right, images of tissue sections showing a physiological effect through labelling the protein corresponding to the centre gene of the subnetwork.

Subnetwork nodes are coloured by differential expression p-value, indicating change of the corresponding transcripts in CCM2-perturbed mice, relative to controls. The border colour of each node indicates transcript fold change. The subnetwork, centreed around the Van Willebrand Factor (VFW) gene, had been identified from the transcriptomic data by the LEAN method[11] to identify regions of strong aggregate differential expression p-value. The aggregates in the image on the right-hand side of the figure show a dysfunction of the VWF pathway in a cerebral tissue section of the CCM mouse model. The observed dysfunction may play a role in the pathophysiology of the disease. For details, we refer the reader to the original publication.[11]

## 3.1.4 Discussion and Perspectives
*Towards Comprehensive Understanding of Complex Disease*

Diseases, such as cancer, are rarely interpretable as the consequence of an abnormality in a single gene. Concepts and tools around functional interaction networks are still evolving, but it is clear that networks offer an important basis for interpreting complex disease, and treating it in the future.[14] For instance, functional interaction networks may be helpful in helping understand the otherwise often cryptic sets of similarities and differences between related diseases. The reason for this may simply be that the sets of genes that play important roles in related diseases strongly

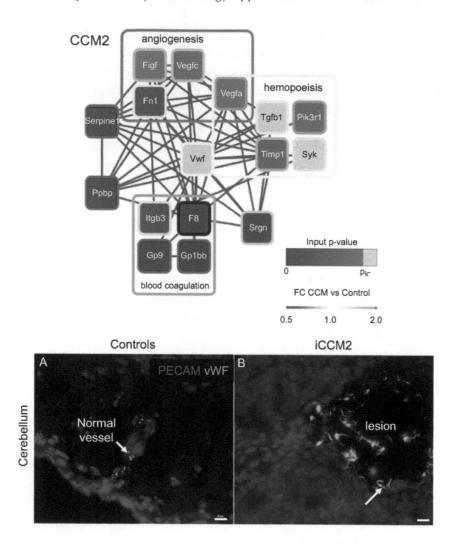

FIGURE 3.1 Top: Cytoscape visualization of a functional subnetwork. Bottom: Physiological effect discovered by fluorescent labelling of the protein in the centre of the subnetwork.[11]

overlap (e.g. Smoller et al.[15]). Thus, understanding disease at the level of functional interaction networks may be essential to develop drug targets and biomarkers, and to reposition existing drugs.

As technologies for molecular profiling continue to proliferate, functional interaction networks for biology and medicine will become more complete, and thus more powerful. In parallel, better and more comprehensive data at the molecular level will also lead to better models. Novel technologies to profile single cells at the molecular level are starting to provide sufficient measurement precision to deduce regulatory networks.[16] Functional interaction networks are thus poised to play an increasingly important role for the development of predictive and preventive approaches to medicine.

## References

1. Tomczak, K., Czerwińska, P. & Wiznerowicz, M. The Cancer Genome Atlas (TCGA): An immeasurable source of knowledge. *Wspolczesna Onkol.* **1A**, A68–A77 (2015).
2. Hanahan, D. & Weinberg, R.A. Hallmarks of cancer: The next generation. *Cell* **144**, 646–674 (2011).
3. Martignetti, L., Calzone, L., Bonnet, E., Barillot, E. & Zinovyev, A. ROMA: Representation and quantification of module activity from target expression data. *Front. Genet.* **7**, 1–12 (2016).
4. Kanehisa, M., Sato, Y., Kawashima, M., Furumichi, M. & Tanabe, M. KEGG as a reference resource for gene and protein annotation. *Nucleic Acids Res.* **44**, D457–D462 (2016).
5. Fabregat, A. et al. The reactome pathway knowledgebase. *Nucleic Acids Res.* **46**, D649–D655 (2018).
6. Szklarczyk, D. et al. STRING v11: Protein–protein association networks with increased coverage, supporting functional discovery in genome-wide experimental datasets. *Nucleic Acids Res.* **47**, 1–7 (2018).
7. Oughtred, R. et al. The BioGRID interaction database: 2019 update. *Nucleic Acids Res.* **47**, D529–D541 (2019).
8. Shannon, P. et al. Cytoscape: A software environment for integrated models of biomolecular interaction networks. *Genome Res.* **13**, 2498–2504 (2003).
9. Cline, M.S. et al. Integration of biological networks and gene expression data using cytoscape. *Nat. Protoc.* **2**, 2366–2382 (2007).
10. Lotia, S., Montojo, J., Dong, Y., Bader, G.D. & Pico, A.R. Cytoscape app store. *Bioinformatics* **29**, 1350–1351 (2013).
11. Gwinner, F. et al. Network-based analysis of omics data: The LEAN method. *Bioinformatics* **33**, 701–709 (2017).
12. Bergametti, F. et al. Mutations within the programmed cell death 10 gene cause cerebral cavernous malformations. *Am. J. Hum. Genet.* **76**, 42–51 (2005).
13. Boulday, G. et al. Developmental timing of CCM2 loss influences cerebral cavernous malformations in mice. *J. Exp. Med.* **208**, 1835–1847 (2011).
14. Loscalzo, J., Barabási, A.-L. & Silverman, K.E. *Network Medicine: Complex Systems in Human Disease and Therapeutics* (Harvard University Press, Cambridge, MA, 2017).
15. Smoller, J.W. et al. Identification of risk loci with shared effects on five major psychiatric disorders: A genome-wide analysis. *Lancet* **381**, 1371–1379 (2013).
16. Rubin, A.J. et al. Coupled single-cell CRISPR screening and epigenomic profiling reveals causal gene regulatory networks. *Cell* **176**, 361.e17–376.e17 (2019).

## 3.2 DISEASE PERCEPTION: PERSONALIZED COMORBIDITY EXPLORATION

*Alfonso Valencia and Jon Sánchez-Valle*

### 3.2.1 Summary

Comorbidity is an impactful medical problem that is attracting increasing attention in healthcare and biomedical research. However, little is known about the molecular processes leading to the development of a specific disease in patients affected by other conditions. We present the *Disease PERCEPTION system*, a web-based tool to analyze the molecular bases of comorbidity relations between 135 diseases and phenotypes. Three levels of information are contained in the system: First, a layer containing the Diseases' Molecular Similarity Network, which provides positive and negative relative molecular similarity interactions between diseases based on the molecular similarity of the patients, significantly recapitulating epidemiologically documented comorbidities. Second, a layer containing the Stratified Comorbidity Network, a network composed of patient-subgroups, conformed by patients with similar molecular profiles, connected by their relative molecular similarities. Third, an additional layer of drugs and small molecules superimposed on the Stratified Comorbidity Network, representing potentially related treatments and drug repurposing options. *Disease PERCEPTION* is designed to provide hypotheses on the molecular bases of previously observed disease comorbidity relations.

### 3.2.2 Introduction

Comorbidity is defined as the presence of one or more secondary diseases associated to a primary disease. The incidence of such comorbidities increases with age, having a high impact on life expectancy, which decreases considerably in the presence of a handful of simultaneous diseases,[1] as observed in ageing populations.[2] Complementarily to direct comorbidity relations, there are also diseases that protect against the development of specific secondary diseases, also known as inverse comorbidity. One of the most renowned examples of inverse comorbidity relations is the one described between central nervous system disorders and cancer.[3]

Assuming a potential role of genetics in the comorbidity relations (as denoted by the lower than expected probability of developing cancer in schizophrenic patients' relatives[4]), several studies have analyzed the molecular bases of comorbidity relations between Alzheimer's disease and cancer. Directly comorbid Alzheimer's disease and glioblastoma present similar differential expression profiles, pointing to the establishment of a chronic inflammatory state in the brain as the molecular bond of both diseases. In the case of inversely comorbid diseases (Alzheimer's disease and lung cancer), their genes are oppositely regulated, specifically the ones involved in the regulation of mitochondrial metabolism, considered as a potential driver of such relation.

Despite these general observations, not all the patients with central nervous diseases are protected against the development of cancer. As an example, in the case of schizophrenia, such relations differ depending on the gender of the patients, protecting males and increasing the risk in females,[4] as well as on the age of onset and duration of the disease (overall, cancer risk varies inversely with age at diagnosis and disease duration).[5] Such results point to the necessity of patient-stratification for the analysis of comorbidities. Therefore, we decided to stratify patients based on their molecular similarities and to explore the molecular bases of their comorbidity relations.

To this end, we calculated patients' molecular similarities based on their differential expression profiles, and used them to calculate relative molecular similarity interactions (interpreted as relative risk interactions) between diseases (which significantly matched previously generated epidemiological networks). Then, for each disease, we extracted patient-subgroups with shared molecular alterations, estimated their relative molecular similarities and identified genes/pathways/drugs potentially driving such relations.[6] To this network of disease subgroups, we add information on potentially related drugs/chemicals with potentially similar expression profiles. We made freely available all the obtained results on the 'Disease PERCEPTION portal'.

### 3.2.3 Approach and Application Example

Disease PERCEPTION stores all the results obtained by our previous study.[6] In summary, it contains information for 6,284 patients suffering from 135 diseases and phenotypes (including smoking and aging). Additionally, each patient is associated to a specific subgroup of molecularly similar patients suffering from the same disease (based on their differential expression profiles). Each of the 1,306 patient-subgroups are associated to genes that are differentially expressed in the same direction in all the patients composing the subgroup. Furthermore, based on the LINCS[7] analysis, patients are positively/negatively connected to drugs if the gene expression changes generated by the drugs are significantly similar/opposite to the ones observed in the patient. Such information has been used to associate drugs to patient-subgroups as in the case of the genes, allowing direct and inverse associations of drugs with disease subgroups.

As an example of the capacity of the system to facilitate the exploration of complex comorbidity relationships, we present the case of asthma, chronic obstructive pulmonary disease (COPD), diabetes, schizophrenia and Alzheimer's disease, all of them previously related by epidemiological studies (Figure 3.2A). The visualization in the *Disease PERCEPTION*

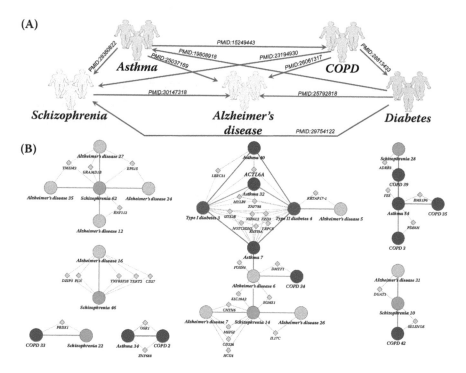

FIGURE 3.2   Case study. (A) Epidemiologically described comorbidity interactions between asthma, chronic obstructive pulmonary disease (COPD), schizophrenia, Alzheimer's disease and diabetes. Blue edges represent direct comorbidity relations. (B) Stratified Comorbidity Network. Circles denote patient-subgroups with at least four patients. Diamonds denote genes. Solid blue and red lines indicate positive and negative relative molecular similarity interactions respectively (evidences of direct and inverse comorbidities). Dashed blue and red lines indicate that the gene is up- or down-regulated in all the patients composing the subgroup to which it is connected. Light blue, green, red and brown nodes represent 'Nervous System diseases', 'Mental disorders', 'Respiratory diseases' and 'Metabolic diseases' respectively.

*portal* shows the proposed relations at the level of the Diseases' Molecular Similarity Network, which contains all the positive and negative relative molecular similarity (RMS) interactions obtained between the studied diseases based on the similarities of the corresponding patients' molecular profiles. Among the ten comorbidity relations described at an epidemiological level, the Diseases' Molecular Similarity Network recovers the comorbidities between both schizophrenia and diabetes with Alzheimer's disease. Interestingly, three interactions contradict epidemiological tendencies (asthma COPD, asthma diabetes and diabetes schizophrenia). Since the RMS estimates takes into account – and so it's affected by – the whole universe of diseases under study (and the proportion of each of them), it's expected to lose some of the relations. As an example, our study analyzes 172 and 189 patients suffering from schizophrenia and Alzheimer's disease respectively, while in the Phenotypic Disease Network[8] there are four times more patients suffering from Alzheimer's disease than schizophrenia. Additionally, the number of patients associated to each disease is considerably smaller than the ones used by epidemiological studies, and so, they might not be representative of the most common endotypes of the diseases.

These results have to be interpreted in light of the known molecular heterogeneity of the diseases, which is obvious for diseases in which diagnosis is particularly difficult.[6] Therefore, to better understand these initial comorbidity hypotheses, we analyze the diseases at the level of disease subgroups, in the so-called Stratified Comorbidity Network (SCN). The SCN represents the positive and negative RMS interactions between patient-subgroups, generated based on the clustering of the patients' molecular similarities (in this case, considering patient-subgroups with at least four similar patients). Interestingly, it let us see how the subgroups from the same disease interact with each other, making it easy to detect the subgroups of patients with differential expression profiles opposite to the most common ones. As an example, we can see that Alzheimer's disease subgroups are the ones with less negative RMS interactions (0.6%), while schizophrenia and COPD are the ones with the highest proportions of negative interactions (~5%). Such results were expected based on the previously mentioned molecular heterogeneity of the diseases. Interestingly, the schizophrenia patient-subgroup presenting more intra-disease negative RMS interactions is the only one presenting inverse comorbidity relations with Alzheimer's disease's patient-subgroups, what might be interpreted as a potential hypothesis related to the specific characteristics of that subgroup of patients that could be followed by further exploration.

In order to generate additional molecular hypotheses that could explain the known epidemiologically described comorbidities, we can use the *Disease PERCEPTION* system to filter RMS interactions with associated genes. Negative interactions are selected only if there is at least a gene that is positively connected to one subgroup – meaning that is up-regulated in all the patients composing it – and negatively connected to the other one (the gene is down-regulated in all the patients composing the subgroup). On the other hand, positive interactions are selected only if there is at least a gene that is positively or negatively connected to both subgroups. As shown in Figure 3.2B, we obtain direct comorbidity relations between both diabetes and schizophrenia with Alzheimer's disease (1 and 9 interactions respectively). Additionally, we recover three direct comorbidity interactions between COPD and schizophrenia, one between COPD and Alzheimer's disease, two between asthma and COPD and two between asthma and diabetes. At the same time, we still detect two inverse comorbidity relations between asthma and COPD, and four between asthma and diabetes (opposite to what was expected based on epidemiological studies). Interestingly, the gene ACTL6A, which plays a role in the proteolysis, is detected to be up-regulated in the type II diabetes subgroup 4, and at the same time to be down-regulated in the asthma subgroup 40, potentially playing a role in the negative relation between both subgroups. Remarkably, protein degradation has been previously described to be increased both in insulin deficient and insulin resistant humans.[9] Regarding the asthma subgroup, a differential proteolytic activity between eosinophilic and neutrophilic asthma has been previously described,[10] potentially explaining the unexpected comorbidity relation. On the other hand, directly comorbid asthma subgroup 32 and type II diabetes 4 share the gene MYLPF deregulated in the same direction, which has been previously described to be associated both to asthma[11] and diabetes.[12] As another interesting case, COPD subgroup 39 and schizophrenia subgroup 28 are detected to be directly comorbid, both of them presenting the gene ADRB3 to be down-regulated. Interestingly, polymorphisms affecting adrenergic receptors have been associated to COPD in adults,[13] and at the same time, several papers provide indirect evidences that adrenergic receptors may play an important role in schizophrenia.[14] Finally, directly comorbid Alzheimer's disease subgroup 7 and schizophrenia subgroup 14 share, among others, the CNTN6 gene as down-regulated. Remarkably, deletions of the entire CNTN6 gene have been previously described in neuropsychiatric disorders (including schizophrenia),[15] while duplications have been detected at 3p26.3, disrupting the CNTN6 gene in Alzheimer's disease.[16]

As mentioned in the introduction, the SCN can be also filtered by shared drugs, that is, drugs that can be associated to patient-subgroups by the similarities of their expression profiles, and might at the same time be involved in the comorbidity relations between patient-subgroups (as previously done with the genes). Suramin, a drug initially used to treat the human African trypanosomiasis or sleeping sickness, caused by *Trypanosoma brucei rhodesiense* and *Trypanosoma brucei gambiense*,[17] was found to be negatively associated both to schizophrenia subgroup 14 and Alzheimer's disease subgroup 6. Interestingly, it has been observed that a single dose of Suramin in the maternal immune activation mouse model of neuropsychiatric disorders restores normal social behavior.[18] At the same time, it has been demonstrated that HDAC inhibitors (as Suramin) constrain tau protein hyperphosphorylation and it is suggested that they enhance synaptic plasticity and learning in neurodegenerative disorders as Alzheimer's disease.[19] These results, together with the ones described in the paper regarding Alzheimer's disease and non-small cell lung cancer,[6] denote the capacity of the developed methodology to generate molecular hypotheses that could be on the basis of disease comorbidities.

### 3.2.4 Discussion and Perspectives

Disease PERCEPTION is an ongoing initiative to help biomedical researchers to generate and/or verify hypotheses on the possible molecular basis of comorbidity relationships, or to explore potentially related drugs that can be involved in the comorbidity relations of their diseases of interest. Future extensions of the system will include:

- Adapt the *Disease PERCEPTION* methodology to other genomic technologies. The methodology will have to be adapted and tested to expand the number of analyzed diseases, as well as to increase the number of patients under study.

- Develop methods to classify new patients into their corresponding diseases and disease subgroups, providing the user with a list of the most probable diseases and their potential molecular drivers.

- Allow the search of the genes of interest, in order to retrieve not only the profile of diseases where the gene is involved, but also the comorbidity relations where those genes could be key players.

- Validate disease-gene associations with information mined from biomedical text (EHR and others).

- Indicate if the comorbidity relations shown in the portal have been previously described by epidemiological and medical studies (including studies analyzing inverse comorbidities).

- Add information regarding the most commonly used drugs to treat each disease, helping to prioritize drugs on the basis of the comorbidity risks associated to the patients' molecular profiles.

## Useful Resources

*Disease PERCEPTION portal:* https://life.bsc.es/compbio/disease-perc eption/

## References

1. DuGoff, E.H., Canudas-Romo, V., Buttorff, C., Leff, B. & Anderson, G.F. Multiple chronic conditions and life expectancy. *Med. Care* **52**, 688–694 (2014).
2. Barnett, K. et al. Epidemiology of multimorbidity and implications for health care, research, and medical education: A cross-sectional study. *Lancet* **380**, 37–43 (2012).
3. Catalá-López, F. et al. Inverse and direct cancer comorbidity in people with central nervous system disorders: A meta-analysis of cancer incidence in 577,013 participants of 50 observational studies. *Psychother. Psychosom.* **83**, 89–105 (2014).
4. Ji, J. et al. Incidence of cancer in patients with schizophrenia and their first-degree relatives: A population-based study in Sweden. *Schizophr. Bull.* **39**, 527–536 (2013).
5. Lin, G.-M. et al. Cancer incidence in patients with schizophrenia or bipolar disorder: A nationwide population-based study in Taiwan, 1997–2009. *Schizophr. Bull.* **39**, 407–416 (2013).
6. Sanchez-Valle, J. et al. Unveiling the molecular basis of disease co-occurrence: Towards personalized comorbidity profiles. *bioRxiv* 431312 (2018).
7. Subramanian, A. et al. A next generation connectivity map: L1000 platform and the first 1,000,000 profiles. *Cell* **171**, 1437.e17–1452.e17 (2017).
8. Hidalgo, C.A., Blumm, N., Barabási, A.-L. & Christakis, N.A. A dynamic network approach for the study of human phenotypes. *PLoS Comput. Biol.* **5**, e1000353 (2009).
9. Hu, J., Klein, J.D., Du, J. & Wang, X.H. Cardiac muscle protein catabolism in diabetes mellitus: Activation of the ubiquitin-proteasome system by insulin deficiency. *Endocrinology* **149**, 5384–5390 (2008).

10. Simpson, J.L., Scott, R.J., Boyle, M.J. & Gibson, P.G. Differential proteolytic enzyme activity in eosinophilic and neutrophilic asthma. *Am. J. Respir. Crit. Care Med.* **172**, 559–565 (2005).
11. Jevnikar, Z. et al. Epithelial IL-6 trans-signalling defines a new asthma phenotype with increased airway inflammation. *J. Allergy Clin. Immunol.* **14**, 577–59 (2018).
12. Giebelstein, J. et al. The proteomic signature of insulin-resistant human skeletal muscle reveals increased glycolytic and decreased mitochondrial enzymes. *Diabetologia* **55**, 1114–1127 (2012).
13. Matheson, M. C. et al. β2-adrenergic receptor polymorphisms are associated with asthma and COPD in adults. *J. Hum. Genet.* **51**, 943–951 (2006).
14. Maletic, V., Eramo, A., Gwin, K., Offord, S.J. & Duffy, R.A. The role of norepinephrine and its α-adrenergic receptors in the pathophysiology and treatment of major depressive disorder and schizophrenia: A systematic review. *Front. Psychiatry* **8**, 42 (2017).
15. Hu, J. et al. CNTN6 copy number variations in 14 patients: A possible candidate gene for neurodevelopmental and neuropsychiatric disorders. *J. Neurodev. Disord.* **7**, 26 (2015).
16. Ghani, M. et al. Genome-wide survey of large rare copy number variants in alzheimer's disease among caribbean hispanics. *G3* **2**, 71–78 (2012).
17. Steverding, D. The development of drugs for treatment of sleeping sickness: A historical review. *Parasit. Vectors* **3**, 15 (2010).
18. Naviaux, J.C. et al. Reversal of autism-like behaviors and metabolism in adult mice with single-dose antipurinergic therapy. *Transl. Psychiatry* **4**, e400 (2014).
19. Xu, K., Dai, X.-L., Huang, H.-C & Jiang, Z.-F. Targeting HDACs: A promising therapy for alzheimer's disease. *Oxid. Med. Cell. Longev.* **2011**, 1–5 (2011).

## 3.3 DECONVOLUTION OF HETEROGENEOUS CANCER OMICS DATA

*Urszula Czerwinska, Nicolas Sompairac and Andrei Zinovyev*

### 3.3.1 Summary

One of the challenges tackled by the computational systems biology of cancer consists of determining the cellular composition of a bulk tumoural sample, where cells of different types, including tumoural cells, are mixed together and collectively contribute to the measured molecular profiles. This problem, called deconvolution of molecular profiles in this context, has attracted a lot of attention with the development of multiple computational approaches in recent years. It is believed that detailed quantification of the cellular composition of existing large-scale cohorts of bulk tumoural samples should give a clue to the success determinants of the application of cancer therapies, as it sheds more light on the possible modulations of the intrinsic immune system response which is often a key factor of patient survival and the response to the therapy. In this short focus chapter, first, we review the algorithmic principles underlying the majority of existing approaches for deconvoluting transcriptomic tumoural profiles and list the major difficulties met in the application of those algorithms in practice. Secondly, we highlight that a family of methods for unsupervised deconvolution remains under-explored in the field. We suggest that this approach, properly used, can lead to the improved definition of context-specific molecular profiles of cell types present in the tumoural microenvironment. Lastly, we comment on some preliminary results obtained from a pipeline we developed, DeconICA, for blind deconvolution of molecular profiles which we tested on a large corpus of publicly available transcriptomic data.

### 3.3.2 Introduction

Tumours are engulfed in a complex microenvironment (TME) including tumour cells, fibroblasts and diversity of immune cells. Understanding the composition of TME in each tumour case is critically important to make a prognosis on the tumour progression and its response to treatment (especially, those treatments based on modulating the immune system response).[1,2] One part of this challenge is to be able to quantify the cellular composition of a tumour sample (called deconvolution problem in this context), using its bulk omics profile (global quantitative profiling of certain types of molecules, such as mRNA or epigenetic markers).[3] In recent years, there has been a remarkable explosion in the number of methods,[4] approaching this problem in several different ways. In this short chapter, we will mainly focus on deconvoluting transcriptomics profiles.

The most general distinction between computational deconvolution approaches is use of the assumption that references molecular profiles of purified cell subpopulations. If the set of reference profiles is known a priori then one can call the deconvolution task supervised or deterministic.[5] In this case only the proportions of the profiles present in the mixtures are estimated. If a method aims to simultaneously infer the cellular composition of a sample and estimate the molecular profiles of the components, then it can be called blind or unsupervised.[6–8] The unsupervised approach represents obviously a more complex task. It can be performed only if a sufficient number of samples, characterized by sufficient variance in cellular composition, is included in the analysis.[9]

In both cases, the bulk molecular profile is supposed to be a weighted linear combination of reference profiles, where weights reflect relative or absolute abundances of the corresponding components. Moreover, both reference profiles and weights are naturally assumed to be described by non-negative numbers. Therefore, natural machine learning approaches for the deconvolution problem are non-negative regression[10] and matrix factorization with non-negative constraints. Since the symbolic birth of the field in 2001[11] many approaches of cell-type deconvolution are being proposed (Figure 3.3A). Indeed, most used algorithms for solving supervised and blind deconvolution problems rely on a form of non-negative least squares regression (NNLS), probabilistic modelling with the assumption of non-negativity (Bayesian approaches) or non-negative matrix factorization (NMF) methods (Figure 3.3B).

### 3.3.3 Approach and Application Example

Direct application of the regression and matrix factorization approaches to solve the problem of tumour bulk sample deconvolution meets certain difficulties in practice, which can be characterized as of a mathematical and biological nature. From the mathematical point of view, one major obstacle for application of deconvolution is in the observation that transcriptomic profiles are roughly characterized by log-normal value distributions. In this situation, the application of least-square minimisation-based methods (such as NNLS or NMF) becomes suboptimal and numerically unstable.

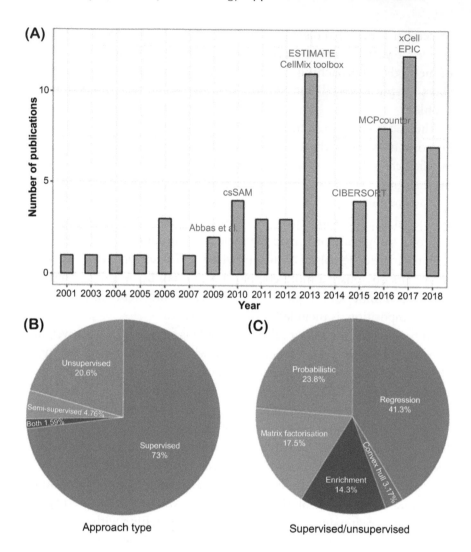

FIGURE 3.3 The overview of number and type of cell-type deconvolution approaches. Despite the growing number of publications, the most popular type of approach remains supervised type of algorithms with regression algorithms as the most popular one. (A) The evolution of the number of publications of cell-type deconvolution algorithms. Stepping stone publications are mentioned in red. (B) Proportion of different mathematical approaches for inferring cell-type proportions (regression, probabilistic, matrix factorization, enrichment and convex hull). (C) Type of the approach (supervised, semi-supervised, unsupervised or both).

The usual trick of taking the logarithm in order to make the distribution close to a bell-shape is not possible to apply because it violates the main assumption of the deconvolution problem (mixture linearity). The second major complication of direct application of regression-based methods for solving the deconvolution problem is in the collection of reference profiles. They are frequently highly correlated which leads to the numerical problems when inverting the mixture matrix in regression. This is because the cell type and subtype differences in regard to the molecular profile are not always quantitatively sufficient to be picked up by an algorithm. Finally, strictly speaking, applying regression is possible only if the set of reference profiles is comprehensive: Meaning that all components of the mixture must be known in advance and not strongly collinear.

From the biological side, a major difficulty consists of the knowledge of reference profiles which are frequently collected from unrelated contexts (such as peripheral blood)[12] and using technological platforms not matching the analyzed datasets. In addition, bulk transcriptomics profiles can be affected by factors transversal to several cell types (such as proliferation),[13] the effect of which can also be negative (inhibition of certain genes). Finally, one of the major components of the tumoural ecosystem – tumoural cells themselves – cannot be characterized in terms of a single reference profile because of their heterogeneity, which makes the regression problem underdetermined.

In the absence of golden standards for evaluating the performance of computational approaches for solving the deconvolution problem, several pragmatical tricks have been exploited in order to deal with the mentioned difficulties. Thus, in different works, it has been suggested to use regularized regression in order to reduce the over-fitting effects, use non-least square minimisation problems, relax the non-negativity constraint or linearity assumption. Also, to deal with the profile collinearities and selection of the most 'distinctive genes', an optimisation of condition number matrix was proposed[10] as well as a feature selection.[12] Some successful methods (such as MCPCounter[14] and xCell[15]) abandon all the regression-based problem formulation and rely on more or less simple scoring approaches (such as taking the mean expression of marker genes or computing enrichment scores) in order to compute some relative cellular component abundances (which are comparable between tumour samples but not comparable between different cellular components and between datasets). At the same time, certain progress in deconvolution has been made through the use of context-specific reference profiles derived with the use of single-cell RNA-Seq approaches[16,17] and/or identifying highly specific biomarkers for particular cell types.[14]

In light of the development of supervised deconvolution approaches, we believe that the potential of the application of blind deconvolution methods remains under-explored (Figure 3.3C). Indeed, a successful blind deconvolution method would bring advantages in solving the deconvolution problem. Among these advantages, there is the possibility to estimate realistically the detectable number of cell types and a possibility to establish an absence of certain cellular components in a bulk sample (which can be interesting information by itself). Unsupervised deconvolution techniques can derive signals not directly related to cell abundance (such as proliferation or technical biases not known in advance). They can also provide an added value as meta-analysis of several datasets through mapping factors common to several datasets on each other.[9] Those methods can also be easier to adapt than supervised techniques, to other complex cell mixtures such as brain cells or tissue mixtures.

Having this motivation in mind, we recently exploited several commonly used matrix factorization methods for their potential to work in a computational pipeline for unsupervised deconvolution of tumour bulk transcriptomic profiles. We observed that use of Stabilized Independent Component Analysis[18,19] accompanied by a method for identifying the optimal number of components[20] usually leads to the *de novo* identification of reference profiles of cell types existing in TME. These discovered cell profiles came out to be more robust across different data sources when compared to more classical Non-Negative Matrix Factorization or Principal Component Analysis.[21,22]

We implemented a pipeline for blind bulk molecular profile deconvolution as an R package named 'DeconICA'.[23] By applying this software to the standard benchmark datasets (simulated and blood transcriptome), it was demonstrated that DeconICA is able to quantify immune cells with accuracy comparable to published state-of-the-art supervised methods but without *a priori* defining cell type-specific signature genes. The implementation can work with existing deconvolution methods based on matrix factorization techniques such as Independent Component Analysis (ICA) or Non-Negative Matrix Factorization (NMF) or most of the other methods that generate putative cell profiles. The proposed pipeline brings more visibility and interpretability to the generated components through correlation with reference profiles and gene enrichment analysis. It proposes high flexibility of extending the default base of reference profiles with user-defined profiles and integration of different matrix factorization algorithms. It also computes the cell abundance scores that are linearly related with the true cell type absolute abundance.

DeconICA, available as an R package from github (https://github.com/UrszulaCzerwinska/DeconICA) was applied to a big corpus of data containing more than 100 transcriptomic datasets composed of, in total, over 28,000 samples of 40 tumour types generated by different technologies and processed independently. This analysis demonstrated that ICA-based immune signals are reproducible between datasets. The three major immune cell types, T-cells, B-cells and Myeloid cells, can be reliably identified and quantified in most of the datasets.

Additionally, the ICA-derived metagenes were used as context-specific signatures in order to study the characteristics of immune cells in different tumour types. The analysis revealed a large diversity and plasticity of immune cells dependent and independent on tumour type. Some conclusions of the study can be helpful in the identification of new drug targets or biomarkers for immunotherapy of cancer.

### 3.3.4 Discussion and Perspectives

Reliable and realistic deconvolution of bulk tumour molecular profiles remains a difficult task. Various tools developed for solving this problem frequently collect contradictory or unstable results.[24] One of the major problems in identifying the most efficient solutions in the field of computational deconvolution lies in the absence of golden standard benchmarks.[5] Such a benchmark would include hundreds of the omics bulk samples paired with an independent assessment of the abundance and profiles of the cellular compartments of the TME.

Currently, most used computational approaches to deconvolution exploit approximations of purified cell type-specific molecular profiles, derived from the contexts (such as peripheral blood) which can be too distant from the tumoural microenvironment ecosystem. The computational pipeline DeconICA that we have developed for blind deconvolution of bulk tumoural omic profiles is able to deal with large cohorts of cancer transcriptomes and automatically derive candidate reference molecular profiles and quantify their abundance. Application of this pipeline to a large collection of tumoural samples, combining TCGA and other datasets, allows a better understanding of what the best settings are and the contexts for the use of existing reference stromal and immune cell signatures.

To conclude, we believe that a successful method for deconvoluting bulk molecular profiles should combine the best features and strongest sides of supervised and unsupervised approaches. In the near future, better development of single cell transcriptomics and multi-omic approaches may provide an additional angle through which the deconvolution problem can be seen. On the other hand, the upcoming Tumour Deconvolution Dream Challenge (https://www.synapse.org/#!Synapse:syn15589870/wiki/582446) can attract even greater focus from the scientific community on this problem and result in novel meta solutions. A better calibration of computational deconvolution methods should quickly translate into onco-immunology research progress and therefore more efficient cancer treatments.

## Useful Resources

DeconICA User Guide: https://urszulaczerwinska.github.io/DeconICA/

R Package Testing Selected Recent Deconvolution Algorithms: https://grst.github.io/immunedeconv/

CellMix R Package for Semi-supervised and Older Regression-based Deconvolution:

http://web.cbio.uct.ac.za/~renaud/CRAN/web/CellMix/

## References

1. Quail, D.F. & Joyce, J.A. Microenvironmental regulation of tumor progression and metastasis. *Nat. Med.* **19**, 1423–1437 (2013).
2. Fridman, W.H., Pagès, F., Sautès-Fridman, C. & Galon, J. The immune contexture in human tumors: Impact on clinical outcome. *Nat. Rev. Cancer* **12**, 298–306 (2012).
3. Shen-Orr, S.S. & Gaujoux, R. Computational deconvolution: Extracting cell type-specific information from heterogeneous samples. *Curr. Opin. Immunol.* **25**, 571–578 (2013).
4. Czerwinska, U. *Computational Deconvolution of Transcriptomics Data from Mixed Cell Populations* (Paris Descartes, 2018).
5. Avila Cobos, F., Vandesompele, J., Mestdagh, P. & De Preter, K. Computational deconvolution of transcriptomics data from mixed cell populations. *Bioinformatics* **34**, 1969–1979 (2018).
6. Wang, N. et al. Mathematical modelling of transcriptional heterogeneity identifies novel markers and subpopulations in complex tissues. *Sci. Rep.* **6**, 18909 (2016).
7. Gaujoux, R. & Seoighe, C. Semi-supervised nonnegative matrix factorization for gene expression deconvolution: A case study. *Infect. Genet. Evol.* **12**, 913–921 (2012).

8. Zinovyev, A., Kairov, U., Karpenyuk, T. & Ramanculov, E. Blind source separation methods for deconvolution of complex signals in cancer biology. *Biochem. Biophys. Res. Commun.* **430**, 1182–1187 (2013).

9. Biton, A. et al. Independent component analysis uncovers the landscape of the bladder tumor transcriptome and reveals insights into luminal and basal subtypes. *Cell Rep.* **9**, 1235–1245 (2014).

10. Abbas, A.R., Wolslegel, K., Seshasayee, D., Modrusan, Z. & Clark, H.F. Deconvolution of blood microarray data identifies cellular activation patterns in systemic lupus erythematosus. *PLoS One* **4**, e6098 (2009).

11. Venet, D., Pecasse, F., Maenhaut, C. & Bersini, H. Separation of samples into their constituents using gene expression data. *Bioinformatics* **17** Suppl 1, S279–S287 (2001).

12. Newman, A.M. et al. Robust enumeration of cell subsets from tissue expression profiles. *Nat. Methods* **12**, 453–457 (2015).

13. Kotliar, D. et al. Identifying gene expression programs of cell-type identity and cellular activity with single-cell RNA-seq. *bioRxiv* (2018).

14. Becht, E. et al. Estimating the population abundance of tissue-infiltrating immune and stromal cell populations using gene expression. *Genome Biol.* **17**, 218 (2016).

15. Aran, D., Hu, Z. & Butte, A.J. xCell: Digitally portraying the tissue cellular heterogeneity landscape. *Genome Biol.* **18**, 220 (2017).

16. Racle, J., de Jonge, K., Baumgaertner, P., Speiser, D.E. & Gfeller, D. Simultaneous enumeration of cancer and immune cell types from bulk tumor gene expression data. *Elife* **6**, e26476 (2017).

17. Schelker, M. et al. Estimation of immune cell content in tumor tissue using single-cell RNA-seq data. *Nat. Commun.* **8**, 2032 (2017).

18. Hyvärinen, A. Fast and robust fixed-point algorithms for independent component analysis. *IEEE Trans. Neural Netw.* **10**, 626–634 (1999).

19. Himberg, J. & Hyvarinen, A. Icasso: Software for investigating the reliability of ICA estimates by clustering and visualization. In *2003 IEEE XIII Workshop on Neural Networks for Signal Processing (IEEE Cat. No.03TH8718)* 259–268 (IEEE, 2003).

20. Kairov, U. et al. Determining the optimal number of independent components for reproducible transcriptomic data analysis. *BMC Genom.* **18**, 712 (2017).

21. Cantini, L. et al. Assessing reproducibility of matrix factorization methods in independent transcriptomes. *Bioinformatics* (2019). doi:10.1093/bioinformatics/btz225.

22. Czerwinska, U., Cantini, L., Kairov, U., Barillot, E. & Zinovyev, A. Application of independent component analysis to tumor transcriptomes reveals specific and reproducible immune-related signals. In 501–513 (Springer, Cham, Switzerland, 2018). doi:10.1007/978-3-319-93764-9_46.

23. Czerwinska, U. UrszulaCzerwinska/DeconICA: DeconICA first release (2018). doi:10.5281/ZENODO.1250070.

24. Sturm, G. et al. Comprehensive evaluation of cell-type quantification methods for immuno-oncology. *bioRxiv* (2018). doi.org/10.1101/463828.

## CHAPTER 4

# Mathematical Modelling of Signalling Networks in Cancer

## 4.1 QUALITATIVE DYNAMICAL MODELLING OF T-HELPER CELL DIFFERENTIATION AND REPROGRAMMING

*Céline Hernandez, Aurélien Naldi and Denis Thieffry*

### 4.1.1 Summary

The balance between different T-helper subtypes, in particular between Th17 and Treg, has been associated with central processes linked to anti-tumoural response. In this chapter, taking advantage of a previously published logical model of T-helper cell differentiation, we explore the possibilities to reach these lineages from naïve cells and to reprogram differentiated cells into alternative phenotypes. For this, we combine several powerful formal approaches and software tools to analyze a logical model containing a hundred components. To ease the recourse to different tools and ensure reproducibility, we have gathered the relevant tools in a *Docker* image, and chained all the analyses in an executable and editable *Python Jupyter* notebook. This notebook covers the computation of asymptotic cell behaviour (stable states), formal state reachability, along with stochastic simulations. From a biological point of view, focusing on the differentiation of Th17 and Treg subsets, we hereby compute the conditions required and the probability of these T-helper subtypes, and we further predict and analyze the effect of micro-environmental perturbations enabling the reprogramming of immuno-supressive Treg into pro-inflammatory Th17.

### 4.1.2 Introduction

At the cellular level, cancer arises from a combination of (epi)genetic perturbations ultimately leading to uncontrolled cell proliferation and resistance to death signals. However, at a higher scale, the development of cancer also strongly impacts the capacity of the immune system to detect and eliminate cancer cells. Indeed, the prognosis of patients has been shown to correlate strongly with the composition of tumour infiltrating immune cells, including dendritic cells, macrophages, natural killer cells and T lymphocytes. In particular, the dominance of specific T-helper (Th) lymphocyte subtypes has been associated with anti-tumoural or inflammatory responses (Th1, Th17), or, on the contrary, with tumour tolerance (Treg).[1,2]

Differentiation into these different subtypes is controlled by specific TCR activation along with various other signals, including dozens of different cytokines. Specific combinations of transcription factors thereby get activated and control downstream differentiation of Th cells. Th subtypes can be categorized according to the expression of key transcription factors along with that of specific cytokines, which can in turn influence the activation and differentiation of various kinds of immune cells.

Over the last decades, several groups have proposed sophisticated models to decipher the interplay between the different signalling pathways and transcriptional regulatory circuits underlying Th cell differentiation.[3] In this chapter, we rely on a recent logical model to explore strategies to trans-differentiate Th cells in order to foster tumour rejection.[4] This model analysis takes advantage of various software tools to compute relevant attractors (stable states or periodic behaviours), to assess the reachability of specific stable states for relevant combinations of input cytokines, to estimate the probability to reach alternative attractors, and to predict cell reprogramming protocols.

One recurring challenge with such study lies in providing proper means to reproduce all computational results and to ease further extensions.[5] In this respect, we have gathered relevant software tools in a *Docker* image, which can be downloaded and installed on computers using recent Linux, Mac OSX or Windows operating systems. We further deliver an executable and editable workflow in the form of a *Python Jupyter* note-book. This contribution is thus made of two parts: this synthetic chapter along with the companion executable notebook and the associated *Docker* image (see Useful Resources section).

### 4.1.3 Approach and Application Example
*CoLoMoTo Virtual Machine and Jupyter Notebook*
While *in silico* analyses are less prone to experimental noise than in vivo counterparts, ensuring their reproducibility and reusability remains a challenging issue. A growing number of scientific publications provide the underlying source code. Such code can be embedded inside *Jupyter* notebooks (or similar tools), which allow mixing it with documentation and visualization of the output results. However, the publication of scripts or notebooks does not ensure that the software on which these scripts rely actually works. Some software tools are difficult to install or depend on precise combinations of libraries which may be impossible to reproduce on newer computer systems, hindering their future reuse. To overcome this second problem, one can use frozen snapshots of complete sets of soft-ware configurations, in particular *Docker* images.

The CoLoMoTo notebook[5] combines these approaches by providing a common *Python* API for a collection of software tools distributed as a *Docker* image. Using this platform, a notebook file can be reproduced by loading it into the image which was used to construct it. In this chapter, we summarize the main steps of our model analysis, which can be followed and re-executed using the *Jupyter* notebook 'Usecase - Balance of Th17 vs Treg cell populations.ipynb' included in the *CoLoMoTo Docker image* (see Useful Resources section).

*Application to Th Cell Differentiation*
This model study is based on the model previously published by Abou-Jaoudé et al.[4] and accounts for the differentiation of naive T-helper cells into various Th subtypes. Built with *GINsim*,[6] this model includes 101 components (including 21 inputs) and 221 interactions (Figure 4.1). Here, we focus the analyses on the pro-Th17 and pro-Treg polarizing environments (Table 4.1A), by assigning fixed activity levels to all inputs excepting IL2 and IL6, without losing important dynamical properties. Using the Java library *bioLQM*,[7] the values of inputs representing the Antigen-Presenting Cell (APC) and TGFB are fixed to 1, while all other inputs excepting IL2 and IL6 are fixed to 0. All fixed values are then

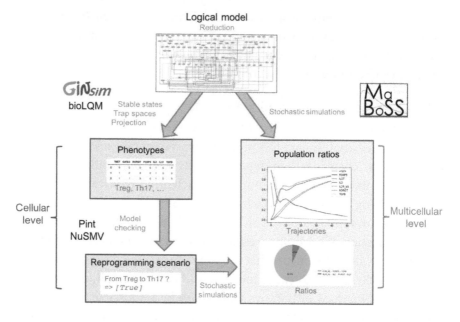

FIGURE 4.1 Depiction of the workflow combining the tools and analyses used in this chapter. A model created with GINsim, can be analyzed from different points of view. At a cellular level, phenotypes can be determined through stable states analysis or trap spaces determination, using GINsim or bioLQM. Next, Pint and NuSMV can be further used to assess attractor reachability and reprogramming scenarios. At a multicellular scale, population ratios can be estimated through stochastic simulations using MaBoSS.

propagated to downstream components and purged from the model. The resulting reduced model contained 31 components, including the two remaining inputs, IL2 for pro-Treg environment, and IL6 for pro-Th17, respectively.

*Computation of Th Cell Asymptotic Behaviours*

The asymptotic behaviours of a model are captured by *attractors*, whether stable or cyclic. They count among the interesting properties of a model representing a biological system, as they can be assimilated to cellular phenotypes. In order to find them, software tools such as *GINsim* or *bioLQM* include efficient methods to formally compute stable states and stable motifs (altogether further corresponding to so-called *trap spaces*), without the need to compute all the state transitions of a model. In the case of our reduced model, we found 12 stable states, along with one additional trap space in the absence of both IL2 and IL6.

TABLE 4.1 Key Properties of Th17 and Treg Lineages and Logical Model Stable States

A

| Lineage | Master Regulator | Secreted Cytokines | Polarizing Environment |
|---|---|---|---|
| Th17 | RORGT | IL17 | APC + TGFB + IL6 |
| Treg | FOXP3 | TGFb | APC + TGFB + IL2 |

B

| IL2_e | IL6_e | TBET | RORGT | FOXP3 | IFNG | IL2 | IL17 | TGFB | Phenotype |
|---|---|---|---|---|---|---|---|---|---|
| 0 | 0 | 0 | 1 | 0 | 0 | 1 | 1 | 0 | Th17 |
| 0 | 0 | 1 | 0 | 0 | 1 | 1 | 0 | 0 | Th1 |
| 0 | 1 | 0 | 1 | 0 | 0 | 1 | 1 | 0 | Th17 |
| 1 | 0 | 0 | 0 | 1 | 0 | 0 | 0 | 1 | Treg |
| 1 | 0 | 1 | 0 | 1 | 0 | 0 | 0 | 1 | Th1-Treg |
| 1 | 0 | 0 | 1 | 0 | 0 | 1 | 1 | 0 | Th17 |
| 1 | 0 | 1 | 0 | 0 | 1 | 1 | 0 | 0 | Th1 |
| 1 | 1 | 0 | 0 | 1 | 0 | 0 | 0 | 1 | Treg |
| 1 | 1 | 0 | 1 | 0 | 0 | 1 | 1 | 0 | Th17 |
| 1 | 1 | 1 | 0 | 1 | 0 | 0 | 0 | 1 | Th1-Treg |

(A) Key properties of Treg and Th17 lineages. Each lineage is characterized by a Master regulator and the production of a specific subset of cytokines. Polarizing environment corresponds to specific signals promoting differentiation into the corresponding lineage. The presence of Antigen Presenting Cells (APCs) is required for all cell activation and differentiation. (B) List of stable states obtained for the reduced model and compatible with pro-Treg and pro-Th17 polarizing environments. Note that a projection is used to show only the most relevant components.

To improve readability, we can further compress this set of attractors by focusing only on a subset of components of interest. For example, by projecting the states onto seven reporter components, TBET, RORGT, FOXP3, IFNG, IL2, TGFB and IL17, we obtain ten stable states (Table 4.1B) and one additional trap space (not shown) representing all the possible asymptotic behaviours for these components. Note that this analysis reports the existence of these attractors but does not provide information regarding the initial conditions required to reach each of them.

### Treg to Th17 Cellular Reprogramming

As mentioned in the previous section, in a cancer context, the balance between Treg and Th17 phenotypes is of particular interest, as Treg cells are mainly associated with tumour tolerance, while Th17 cells are mainly associated with anti-tumoural response.

In a cancer therapy context, it might be interesting to reprogram Treg cells into Th17 cells, in order to promote anti-tumoural response. To identify micro-environmental perturbations enabling such reprogramming, we use the software *Pint*,[8] enabling efficient reachability analyses, in combination with generic model checkers such as *NuSMV*.[9] The Th17 phenotype being the only attractor compatible with a pro-Th17 environment, we thus expect that all cells, in particular Treg cells, must be converted into Th17 cells under this environment. Using *Pint*, we confirm that the reachability of a Th17 phenotype is possible when starting from Treg stable states and changing the cell environment to a pro-Th17 mix of cytokines.

Next, we use the software *MaBoSS*,[10] which enables probabilistic simulations of Boolean models using time Markov processes in combination with the Gillespie algorithm, and verify that all Treg cells are reprogrammed into Th17 cells.

However, cellular environments are rarely so tightly controlled. A mixed environment combining IL2 and IL6 would likely be more realistic. Under this environment, the model gives rise to several attractors, including attractors corresponding to Th17 and Treg phenotypes. Using *Pint*, we show that in this case Treg cells cannot be reprogrammed into Th17 cells simply by adding IL6 to a pre-existing pro-Treg environment but further require the removal of IL2.

As Th17 cells obtained in a pro-Th17 environment do produce IL2, we further hypothesize that IL2 does not prevent the Th17 phenotype by itself. We thus checked if a transient removal of the IL2 response was sufficient to enable reprogramming. As in the previous analysis, we put Treg cells in a mixed environment (i.e. adding both IL2 and IL6) further shutting down the activity of the IL2 receptor in the initial condition. Interestingly, we could thereby verify that reprogramming becomes possible with only a transient blocking of the IL2 signalling pathway.

As a transient shutdown of the IL2 pathway enables reprogramming, we further wondered whether a full inactivation of this pathway could be a good strategy to enforce it. However, when locking the IL2 receptor into a completely inactive state, we lose the capability to convert Treg cells into a Th17 phenotype.

### Analysis at the Population Level

From the preceding analyses, it became clear that Treg cells could be reprogrammed into Th17 cells in a pro-Th17 environment, and we suspected that this outcome is the only possible one in that case. Although the model represents a single cell, the behaviour of a cell population can be studied using the software *MaBoSS*[10] to produce stochastic trajectories over time. These stochastic simulations show that 100% of Treg cells successfully reprogram to adopt a Th17 phenotype when immersed into a pro-Th17 environment. More interestingly, they demonstrate that a smaller fraction (6%) of Treg cells can be reprogrammed into Th17 cells when immersed in a mixed environment containing both IL2 and IL6, but only after a transient inactivation of the IL2 receptor.

### 4.1.4 Discussion and Perspectives

Logical modelling allows the construction of large models, which can then be analyzed with powerful methods and software tools. Current computational strategies to cope with model complexity include model reduction, effective computation of attractors along with their projection on subsets of relevant components. In this chapter, we demonstrate how complementary software tools can be combined in a consistent and reproducible workflow to analyze a model of CD4 T cell differentiation.

Using appropriate tools and methods, we could reduce and simplify a complex model more amenable to dynamical analysis. This enabled us to identify stable states corresponding to Treg and Th17 subtypes, to assess the possibility to reprogram Treg cells into Th17 cells, and even estimate the proportion of successful reprogramming for a population of cells depending on environmental conditions. Although coherent with currently published data, these predictions need to be more challenged by designed specific experiments.

As the whole analysis workflow is provided in the form of a *Python* notebook, it can be easily reproduced or extended after a relatively straightforward installation of the *Docker* software and the corresponding *Docker* image.

## Useful Resources

To run the notebook and the software used in this model analysis, it is necessary to first install *Docker*, along with the *Docker* image 'colomoto/colomoto-docker - version 2018-12-22', available at the URL: https://hub.docker.com/r/colomoto/colomoto-docker. This *Docker* image includes the corresponding *Jupyter* notebook as a use case: 'Usecase - Balance of Th17 vs Treg cell populations.ipynb', which is also available at the URL: https://github.com/colomoto/colomoto-docker/tree/master/usecases.

## References

1. Varn, F.S. et al. Adaptive immunity programmes in breast cancer. *Immunology* **150**, 25–34 (2017).
2. Knochelmann, H.M. et al. When worlds collide: Th17 and Treg cells in cancer and autoimmunity. *Cell. Mol. Immunol.* **15**, 458–469 (2018).
3. Abou-Jaoudé, W. et al. Logical modelling and dynamical analysis of cellular networks. *Front. Genet.* **7**, 94 (2016).
4. Abou-Jaoudé, W. et al. Model checking to assess T-helper cell plasticity. *Front. Bioeng. Biotechnol.* **2**, 86 (2014).
5. Naldi, A. et al. The CoLoMoTo interactive notebook: Accessible and reproducible computational analyses for qualitative biological networks. *Front. Physiol.* **9**, 1–13 (2018).
6. Naldi, A. et al. Logical modelling and analysis of cellular regulatory networks with GINsim 3.0. *Front. Physiol.* **9**, 646 (2018).
7. Naldi, A. BioLQM: A java toolkit for the manipulation and conversion of logical qualitative models of biological networks. *Front. Physiol.* **9**, 1605 (2018).
8. Paulevé, L. in *CMSB 2017 – 15th Conference on Computational Methods for Systems Biology* 309–316 (Springer, 2017).
9. Cimatti, A. et al. NuSMV version 2: An opensource tool for symbolic model checking. In *Proceedings of the International Conference on Computer-Aided Verification (CAV 2002)*, **2404** (Springer, 2002).
10. Stoll, G. et al. MaBoSS 2.0: An environment for stochastic Boolean modelling. *Bioinformatics* **33**, 2226–2228 (2017).

## 4.2 MATHEMATICAL MODELS OF SIGNALLING PATHWAYS AND GENE REGULATION INVOLVED IN CANCER

*Kirsten Thobe, Bente Kofahl and Jana Wolf*

### 4.2.1 Summary

The development of cancer is associated with an accumulation of perturbations in cells, such as mutations or abnormal expression of proteins. Often perturbations affect cellular signalling pathways. These show dynamical responses towards changes in the environment by controlling the gene expression and eliciting appropriate cellular responses. Depending on the type, strength or duration of a stimulus, cellular responses can differ considerably. Perturbations within a cell can strongly affect these cellular responses. In order to understand cancerous behaviour and to develop effective strategies for the treatment of cancer, it is necessary to comprehend the impact of individual and combined perturbations in these highly regulated cellular networks.

We use mathematical modelling and computational analysis to describe cell signalling and to gain a deeper understanding of the regulatory mechanisms. Specifically, we focus on the nuclear factor-κB (NF-κB) signalling pathway which is often dysregulated in tumours but also in other diseases. We explore the effects of different regulatory processes, e.g. feedback regulation, post-transcriptional modifications or cross-talk on the dynamical properties of the pathway and show how quantitative modelling can be employed to identify new regulatory layers. Based on such analyses conclusions about effective invention points for drugs can be drawn.

### 4.2.2 Introduction

Cellular signalling pathways enable a cell to detect and respond to changes in its environment. An appropriate response of a cell towards specific stimuli is the basis for processes such as development, adaptation and immune response. Signalling pathways regulate critical cellular processes, e.g. proliferation, differentiation and apoptosis, mostly by regulating the downstream target gene expression. It was found that the dynamics of signalling components play a pivotal role in the control of gene expression.[1–3]

Many signalling pathways have been described to be dysregulated in cancer cells,[4] for example the p53 pathway, the NF-κB signalling system, the Wnt/β-catenin pathway or the epidermal growth factor receptor (EGFR) signalling. Dysregulation can arise by mutations in proteins, overexpression or reduced expression of pathway components or changes in post-translational modifications and varies between cancer types and individual patients (e.g. The Cancer Genome Atlas[5]). Questions of how signalling pathways, in particular the dynamics of specific pathway readouts, are regulated in normal and perturbed situations have been intensively studied.

Here, we focus on the NF-κB signalling system, which is a pathway critically involved in the development of cancer but also cardiovascular and autoimmune diseases.[6-8] The system comprises a canonical (or classical) pathway and a non-canonical (or alternative) pathway.[9,10] Both signalling branches regulate transcription factors of the NF-κB family controlling a range of gene expression programs that affect inflammatory and immune responses as well as cell proliferation and cell death. In detail, the transcription factors bind the DNA as dimers, where the most common dimer related to the canonical pathway comprises p50 and RelA and the dimer related to the non-canonical pathway comprises p52 and RelB (Figure 4.2). A large number of stimuli can activate the NF-κB pathways regulating a network including receptors, adaptors, ligases and kinases that lead to the occurrence of active NF-κB dimers. The role of the NF-κB dynamics in regulating gene expression has been intensively studied.[11,12]

Numerous mathematical models have been developed to understand the regulatory processes that modulate NF-κB dynamics on different levels of the signalling system.[13-16] In particular, models of the pathway have been used to analyze implications of pathway perturbations, e.g. loss of function of the pathway components IκB or A20.[17,18]

### 4.2.3 Approach and Application Example

We aim to understand the regulation of the NF-κB signalling system using computational modelling (Figure 4.2). In a first project, we focused on the cellular heterogeneity in the dynamics of the canonical pathway. Experimental time-resolved, single cell data have shown that individual cells of a population respond with a variety of NF-κB dynamics to a TNFα stimulus.[11,19] The dynamics range from sustained to damped oscillations, or transient increases and therefore no oscillations of nuclear NF-κB. In experiments based on cell populations[17] damped oscillations were observed. According to the respective experiments, computational models describing either sustained[20] or damped[17,18,21] oscillations had been developed. However, the question remained unsolved as to which intracellular parameters could be the cause of the different dynamical NF-κB responses to identical stimuli. We addressed that question by a bifurcation analysis, which enables the identification of model parameters whose variation leads

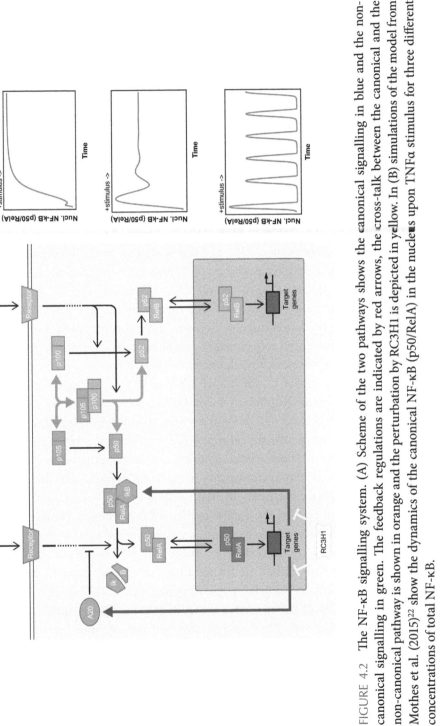

FIGURE 4.2 The NF-κB signalling system. (A) Scheme of the two pathways shows the canonical signalling in blue and the non-canonical signalling in green. The feedback regulations are indicated by red arrows, the cross-talk between the canonical and the non-canonical pathway is shown in orange and the perturbation by RC3H1 is depicted in yellow. In (B) simulations of the model from Mothes et al. (2015)[22] show the dynamics of the canonical NF-κB (p50/RelA) in the nucleus upon TNFα stimulus for three different concentrations of total NF-κB.

to changes in the dynamics. As a first step, we developed a core model of the canonical NF-κB pathway based on an established detailed model that describes damped oscillations of nuclear NF-κB (p50/RelA).[21] Sensitivity analyses for different model outputs have been used to assess the impact of parameters in that detailed model. Processes that only have a small impact on the steady state values and the dynamics were identified and removed from the model. The resulting core model preserved the dynamical and stationary behaviour of NF-κB very well[22] and therefore served as the basis for a bifurcation analysis. This analysis identified two parameters, the total concentration of NF-κB and the transcription rate constant of IκBα, to be critical for the occurrence of damped or sustained oscillations. This finding has been validated in established, more detailed models of the pathway.[22] This implies that a change in each of these identified parameters can change the temporal behaviour of NF-κB from sustained to damped oscillations, or eliminate oscillations as shown in Figure 4.2B for changes in the total NF-κB concentration. Cell-to-cell variation in levels of total NF-κB, e.g. p65, might therefore underlie the different dynamics of nuclear NF-κB measured within a cell population but also between different cell types. Variations of the IκBα transcription rate constant can arise from transcriptional co-factors and might therefore be influenced by cross-talks from other pathways.

The activity of the canonical NF-κB pathway is strongly regulated by negative feedbacks. Important examples are those exerted by the NF-κB target genes IκBα and A20. IκBα is an inhibitor of NF-κB itself, while A20 inhibits upstream processes (Figure 4.2).[18,20,23] The abundance and activity of components of these feedback loops can be modulated by additional regulators which may consequently impact the dynamics of the pathway. In a collaborative experimental-theoretical approach we studied the effect of post-transcriptional regulation by the mRNA-binding protein RC3H1 on the NF-κB system.[24] To that end, a mathematical model of NF-κB signalling including the feedbacks via IκBα and A20 as well as RC3H1 was developed and quantified by fitting to experimental data, in particular, western blot data for A20, IκBα and phosphorylated IKK as well as quantitative PCR (qPCR) data for IκBα and A20 mRNA. This quantitative model can reproduce the temporal behaviour of pathway components under non-perturbed conditions and predict the behaviour of pathway components under conditions of induced or reduced RC3H1 abundances. Quantitative model simulations and experiments showed that RC3H1 affects the expression of the pathway regulators A20 and IκBα and has an impact on the IKK and NF-κB dynamics demonstrating that the NF-κB pathway is affected by post-transcriptional regulation. Thus, RC3H1 constitutes a new tool to manipulate the signalling system.[24]

An additional layer of regulation may arise from possible cross-talks with other signalling pathways or between the two NF-κB signalling branches. This may result from sharing pathway components or by regulating the expression of components of other pathways. For the canonical and the non-canonical NF-κB pathways it is known that they share components such as TRAF2 and IKKα. Further, the canonical pathway regulates the expression of proteins involved in non-canonical signalling, e.g. the precursor protein p100.

The homeostasis and activation of the overall NF-κB signalling system critically depend on the processing of the precursor proteins p100 and p105. Upon activation of upstream receptors, both proteins can be partially degraded by the proteasome resulting in the respective products p52 and p50, which are essential NF-κB transcription factors. While p105 is constitutively expressed and required for canonical signalling, p100 is critically involved in non-canonical signalling but is a target gene of the canonical pathway. To address the question whether the precursors interact or depend on each other via the proteasomal processing, we used a mass-spectrometry based quantitative analysis and a computational modelling approach to dissect the mechanism of NF-κB precursor processing. The applied selected-reaction monitoring (SRM) approach provides an absolute quantification of pathway components. In particular, the total amounts of precursors and their processed products were measured over time with and without non-canonical stimulation. Surprisingly, the experiments revealed that p105 was processed in a non-canonical stimulus dependent manner.[25] In order to explain these findings, four candidate models proposing different mechanistic dependencies between p100 and p105 were developed. A comparison of quantitative model fittings to the time-resolved data showed that p100 and p105 form a complex in resting and stimulated cells that is stimulus responsive. Overall, our investigations showed an interdependent proteolysis of both precursors upon non-canonical stimulation resulting in a concurrent production of p52 and p50. This molecular mechanism establishes a new link between the two NF-κB signalling pathways (see Figure 4.2).

## 4.2.4 Discussion and Perspectives

Computational models are useful tools for exploring biological processes systematically. Model analysis can identify the most sensitive processes and critical regulations, which can be used to predict the effect of perturbations such as mutations but also to find effective drug targets. For the NF-κB system, we used a dynamical model to address the issue of cell-to-cell variation in the dynamical behaviour between cells of a population but also between different cell lines. After developing a core model of canonical NF-κB signalling, we performed a bifurcation analysis that identified two critical parameters for damped or sustained oscillations. This result shows that cell-type specific dynamics for a certain pathway do not necessarily require a different pathway structure, but could be caused by varying intracellular processes. This emphasizes the need for the quantification of cellular parameters to predict cellular processes. Quantitative data combined with modelling also allow for the identification of new mechanisms, as shown for the new regulation layers of the NF-κB system: 1) the RNA binding protein RC3H1 for RNA processing, and 2) the cross-talk between the canonical and non-canonical pathway via p100/p105 precursor processing.

An open problem is the integration of all known components, regulations and processes in comprehensive quantitative pathway models. For that purpose, central intermediates of the system need to be quantified experimentally e.g. using mass-spectrometry derived methods. Quantitative data for different cell types will enable the development of cell-type and condition specific models. For the development of disease-specific models, the integration with other cellular pathways plays an important role as well as effects of the microenvironment, that is, the embedding in a tissue environment. So far, these aspects have only been partly addressed for the NF-κB system, but examples exist for other pathways.[26,27] The integration of diverse biological aspects will require the use of different modelling frameworks, e.g. Boolean approaches or spatially resolved models and possibly hybrid approaches. Thereby, computational models will continue making important contributions to the deeper understanding of pathways and their regulation, which is essential for understanding complex diseases such as cancer and to predict effective strategies for treatment.

## References

1. Purvis, J.E. & Lahav, G. Encoding and decoding cellular information through signalling dynamics. *Cell* **152**(5), 945–956. (2013).
2. Behar, M. & Hoffmann, A. Understanding the temporal codes of intra-cellular signals. *Current Opinion in Genetics & Development* **20**(6), 684–693. (2010).

3. Kolch, W., Halasz, M., Granovskaya, M. & Kholodenko, B.N. The dynamic control of signal transduction networks in cancer cells. *Nature Reviews Cancer* **15**(9), 515. (2015).
4. Hanahan, D. & Weinberg, R.A. Hallmarks of cancer: The next generation. *Cell* **144**, 646–674. (2011).
5. The Cancer Genome Atlas. (2015). Available at https://cancergenome.nih.gov.
6. Hayden, M.S. & Ghosh, S. NF-κB, the first quarter-century: Remarkable progress and outstanding questions. *Genes & Development* **26**(3), 203–234. (2012).
7. Nagel, D., Vincendeau, M., Eitelhuber, A.C. & Krappmann, D. Mechanisms and consequences of constitutive NF-κB activation in B-cell lymphoid malignancies. *Oncogene* **33**(50), 5655. (2014).
8. Ben-Neriah, Y. & Karin, M. Inflammation meets cancer, with NF-κB as the matchmaker. *Nature Immunology* **12**(8), 715–723. (2011).
9. Shih, V.F.S., Tsui, R., Caldwell, A. & Hoffmann, A. A single NF-κB system for both canonical and non-canonical signalling. *Cell Research* **21**(1), 86. (2011).
10. Hinz, M. & Scheidereit, C. The IκB kinase complex in NF-κB regulation and beyond. *EMBO Reports* **15**(1), 46–61. (2014).
11. Nelson, D.E., Ihekwaba, A.E.C., Elliott, M., Johnson, J.R., Gibney, C.A., Foreman, B.E., … & Edwards, S.W. Oscillations in NF-κB signalling control the dynamics of gene expression. *Science* **306**(5696), 704–708. (2004).
12. Zambrano, S., De Toma, I., Piffer, A., Bianchi, M.E. & Agresti, A. NF-κB oscillations translate into functionally related patterns of gene expression. *Elife* **5**, e09100. (2016).
13. Lipniacki, T. & Kimmel, M. Deterministic and stochastic models of NFkappaB pathway. *Cardiovascular Toxicology* **7**(4), 215–234. (2007).
14. Cheong, R., Hoffmann, A. & Levchenko, A. Understanding NF-κB signalling via mathematical modelling. *Molecular Systems Biology* **4**(1), 192. (2008).
15. Basak, S, Behar, M. & Hoffmann, A. Lessons from mathematically modelling the NF-κB pathway. *Immunological Reviews* **246**(1), 221–238. (2012).
16. Williams, R., Timmis, J. & Qwarnstrom, E. Computational models of the NF-KB signalling pathway. *Computation* **2**(4), 131–158. (2014).
17. Hoffmann, A., Levchenko, A., Scott, M.L. & Baltimore, D. The IκB-NF-κB signalling module: Temporal control and selective gene activation. *Science* **298**(5596), 1241–1245. (2002).
18. Lipniacki, T., Paszek, P., Brasier, A.R., Luxon, B. & Kimmel, M. Mathematical model of NF-κB regulatory module. *Journal of Theoretical Biology* **228**(2), 195–215. (2004).
19. Sung, M.H., Salvatore, L., De Lorenzi, R., Indrawan, A., Pasparakis, M., Hager, G.L., … & Agresti, A. Sustained oscillations of NF-κB produce distinct genome scanning and gene expression profiles. *PloS one* **4**(9), e7163. (2009).

20. Ashall, L., Horton, C.A., Nelson, D.E., Paszek, P., Harper, C.V., Sillitoe, K., … & Kell, D.B. Pulsatile stimulation determines timing and specificity of NF-κB-dependent transcription. *Science* **324**(5924), 242–246. (2009).
21. Kearns, J.D., Basak, S., Werner, S.L., Huang, C.S. & Hoffmann, A. IκBε provides negative feedback to control NF-κB oscillations, signalling dynamics, and inflammatory gene expression. *Journal of Cell Biology* **173**(5), 659–664. (2006).
22. Mothes, J., Busse, D., Kofahl, B. & Wolf, J. Sources of dynamic variability in NF-κB signal transduction: A mechanistic model. *BioEssays* **37**(4), 452–462. (2015).
23. Werner, S.L., Kearns, J.D., Zadorozhnaya, V., Lynch, C., O'Dea, E., Boldin, M.P., Ma, A., Baltimore, D. & Hoffmann, A. Encoding NF-kappaB temporal control in response to TNF: Distinct roles for the negative regulators IkappaBalpha and A20. *Genes & Development* **22**(15), 2093–2101. (2008).
24. Murakawa, Y., Hinz, M., Mothes, J., Schuetz, A., Uhl, M., Wyler, E., … & Kempa, S. RC3H1 post-transcriptionally regulates A20 mRNA and modulates the activity of the IKK/NF-κB pathway. *Nature Communications* **6**, 7367. (2015).
25. Yılmaz, Z.B., Kofahl, B., Beaudette, P., Baum, K., Ipenberg, I., Weih, F., … & Scheidereit, C. Quantitative dissection and modelling of the NF-κB p100-p105 module reveals interdependent precursor proteolysis. *Cell Reports* **9**(5), 1756–1769. (2014).
26. Kholodenko, Boris N. Cell-signalling dynamics in time and space. *Nature Reviews Molecular Cell Biology* **7**, 165–176. (2006).
27. Hartung, N., Benary, U., Wolf, J. & Kofahl, B. Paracrine and autocrine regulation of gene expression by Wnt-inhibitor Dickkopf in wild-type and mutant hepatocytes. *BMC Systems Biology* **11**(1), 98. (2017).

## 4.3 DYNAMIC LOGIC MODELS COMPLEMENT MACHINE LEARNING TO IMPROVE CANCER TREATMENT

*Mi Yang, Federica Eduati and Julio Saez-Rodriguez*

### 4.3.1 Summary

Machine learning applied to predicting drug response in cancer cell lines has mainly focused on predicting the efficacy of specific drugs. In this article, we first discuss the application of matrix factorization in cancer drug screenings to learn models for multiple drugs in conjunction. We use the tool Macau to integrate pathway activities inferred with the tool PROGENy as features of cell lines, and targets as features of drugs. This allowed us to explore broader and deeper functional interactions between drug targets and pathway activities. While machine learning is a powerful means to predict drug response and extract global patterns, it can miss specific insights on signalling deregulation in cancer cells that are relevant to understand and improve drug treatment. Therefore, we propose to complement machine learning with more mechanistic approaches, in particular dynamic logic modelling. We describe applications of logic modelling to research on the combinations of cancer drugs, using the tool CellNOpt to train a generic network built with the meta-database of pathways OmniPath, on phospho-proteomic data obtained upon perturbation with ligands and drugs. This contextualizes the generic network into cell-type specific predictive models that we use to find biomarkers and propose combination strategies.

### 4.3.2 Introduction

Precision medicine aims at identifying a subset of patients likely to respond to a specific drug. Identification of the specific population where the efficacy is maximized can increase the chance of success of clinical trials and therefore expand therapeutic options. Combining the wealth of omics data with mathematical analysis is a promising way to identify molecular biomarkers of drug response.

In particular, different machine learning algorithms have been used to predict drug response on large datasets,[1] allowing the integration of different omics profiling datasets (such as genomics, transcriptomics and proteomics) and possibly the selection of relevant features.[2] More mechanistic approaches have also shown great potential in understanding how cellular systems are deregulated in cancer and how drugs can be effectively used to modulate these systems.[3,4]

In order to derive predictive models, a large amount of suitable experimental data is usually required. Treatment data of patients with matching molecular data is still relatively sparse. In contrast, drug screenings on cancer cell lines, complemented with different types of omics data, provide larger datasets to build these predictive models. Cancer cell lines have been the workhorse of preclinical study in oncology drug development and cancer biology research. Their main advantage is that they enable a scalable and controlled study of cellular response to therapies. At the same time, they are a limited model not capturing important aspects of cancer such as the immune response, tumour heterogeneity, the role of microenvironment or vascularity.

In this chapter, we will present some computational tools that we have developed and employed to predict and elucidate the cancer drug response mechanism based on machine learning and mechanistic modelling. Due to space limitations we can not describe other complementary methods from other groups, and we point the reader to recent reviews.[2-4]

### 4.3.3 Approach and Application Example

A hallmark of cancer is the deregulation of signalling pathways. These pathways are often the target of therapies (e.g. antibodies, kinase inhibitors), and their alterations are often associated to drug resistance. Thus, estimation of the level of activation of signalling pathways is useful to understand drugs' efficacy. Pathway activity can be estimated from different omics data. We focus here on transcriptomics, which provides a genome-wide blueprint of the status of the cell, and has been shown to be in general more predictive of drug response than genomics.[1,5] Reducing the high dimensional gene expression data into pathway signatures allows better interpretability, thus a more mechanistic view of the system, and provides more statistical power for further analyses.

Broadly speaking, pathway signatures can be derived by two approaches: from the gene expression of the components of the pathways, or from the expression of the genes downstream of the pathway, that change when the pathway's activity is altered. The latter tend to be more informative[6-8] and, following this paradigm, we have developed *PROGENy*,[6] a data driven method to infer pathway activity scores from high dimensional gene expression data. PROGENy is based on pathway signatures derived from downstream gene expression changes upon perturbation experiments. It uses the footprint on gene-expression of a pathway, rather than at the components of the pathway, to estimate its activity. PROGENy scores are based on many experiments for each pathway to derive robust signatures. It currently covers 14 pathways: EGFR, NFkB, TGFb, MAPK, p53, TNFa, PI3K, VEGF, Hypoxia, Trail, JAK STAT, Androgen, Estrogen and WNT. PROGENy finds more association with survival and drug response than 'mapping methods'.[6] We have also developed *DoRothEA*,[9] a comprehensive collection of knowledge- and data-driven genes target (regulons) for more than 1,500 Transcription Factors (TF). DoRothEA can be used to estimate the activities of TFs from the gene expression level of their regulons using enrichment analysis techniques. Many TFs' activities are correlated with anticancer drugs' response, and combining existing genomic markers with TF activities often improves the drug response stratification of cancer cell lines.[9]

Signatures such as the PROGENy or DoRothEA scores can be fed into a machine learning algorithm in order to predict drug response data, and extract mechanistic insights (Figure 4.3). We use *Macau*,[10] a multi-task learning algorithm based on matrix factorization, that is well suited for cancer drug screenings datasets for several reasons:

1) Prediction of drug sensitivity in a multitask setting improves performance with each task helping the others. In multitask learning, we bring together all drugs in a single model, and learn common patterns reflecting the underlying mechanisms.

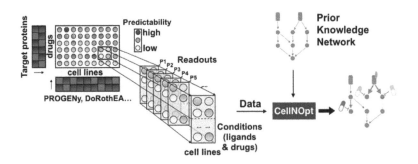

FIGURE 4.3   We use matrix factorization to predict drug response on cancer cell lines. To improve prediction performance, we use information about the drugs, such as protein targets, and information about the cell lines, such as PROGENy scores or DoRothEA transcription factor activities. For cases of low prediction performance, we could focus on those drug-cell line pairs and perform additional perturbation experiments. Analyzing these data with mechanistic modelling, using for example CellNOpt, we can gain additional insights on the mechanism of resistance by feeding those perturbation data with a prior knowledge network to CellNOpt for mechanistic modelling.

2) If we want to predict the response for new drugs, drug features (for example protein target) can be used to improve the prediction performance. This would not be possible in single task learning.

3) It is possible to extract hidden insights from the interactions between the drug features and cell lines features through matrix factorization. This can answer questions such as: Which type of drugs works on which types of cell lines? Using drug target and pathway activity, a potential insight could be: 'Activation of pathway X confers sensitivity to drugs targeting protein Y'. Whereas standard algorithms could only answer the question: Which gene is predictive of which drug?

We have applied Macau[10] for proposing drug repositioning and combination treatments.[7] We used signalling pathways' activities estimated by PROGENy as input features for the cell lines, and manually curated target proteins for the drugs. As application data, we used the Genomics of Drug Sensitivity in Cancer[5] (GDSC), a large collection of around 1,000 molecularly characterized cell lines treated with over 250 drugs.

The prediction performance and biological insights explored were comparable or superior to standard methods such as Elastic Net. Nevertheless, our and other machine learning methods are far from perfect in predicting drug response from these rich yet static omics' data sets. To complement these approaches, a more functional strategy of perturbing cells with drugs and measuring the response can be very informative.[19] This can in particular help to understand the drug's mode of action, and distinguish correlation from causality. This is important as key contributors of drug response predictability can be considered as actionable insights, but they should be supported by a mechanistic rationale.

Since cancer cells are complex dynamic systems, the use of mechanistic models can help in gaining a more refined understanding of the cellular regulatory circuits, therefore helping to address the questions mentioned earlier analysing functional (perturbation-based) data. Information on the components and structure of signalling pathways is currently available from different databases. *OmniPath*[11] is a unified resource of information relevant for signal transduction, encompassing over 50 different databases, including causal interactions, undirected interactions and biochemical reactions. It also integrates additional information on the structure and mechanism of the interactions, drug targets, functional annotation, tissue-specific expression and mutations. Thus, OmniPath provides easy access to a large number of regulatory interactions which can be present in each cell. However, this type of static network cannot be directly used to study context (e.g. tissue, tumour) – specific and dynamic (e.g. response to perturbations) cellular mechanisms.

To convert static generic networks into dynamic specific models we developed *CellNOpt*.[12] In CellNOpt, prior information on the general network topology is trained to perturbation data to build context-specific predictive logic models (Figure 4.3). CellNOpt[13] provides different types of logic formalism, including: (a) Boolean steady state, (b) Boolean multi-time scale, (c) continuous state (Fuzzy logic) and (d) time-resolved logic models based on ordinary differential equations. As an example of application to investigation and prediction of drug effects, we used the latter formalism to understand and target mechanism of drug resistance in 14 colorectal cancer cell lines.[14] We started from a prior knowledge network of the signalling pathways of interest, including 49 nodes (signalling molecules) and 86 edges (regulatory interactions). Cell type-specific logic models were trained using data measuring 14 phosphoproteins upon perturbations with combinations of five stimuli (ligands binding to cell surface, activating downstream pathways) and seven drugs (inhibitors targeting nodes of the network). A regularization technique was applied to improve parameters' identifiability, and the resulting models were then used to study the heterogeneity of the signalling pathway. Additionally, we connected these models to a pharmacogenomic study,[5] looking for associations between mode parameters and drug response. In this way we found a number of model-based biomarkers that were used to successfully predict combinatorial treatments to overcome resistance. In particular we predicted and validated that resistance to MEK inhibitors could be overcome by co-blockade of GSK3 on specific colorectal cancer cell lines.

### 4.3.4 Discussion and Perspectives

While machine learning models have been widely used in cancer research, other approaches rooted in molecular mechanisms can complement them. Our tool CellNOpt allows us to build context-specific predictive logic models of signalling pathways. The models are trained from perturbation data using a prior knowledge network provided by OmniPath. The scope of the models is limited by the readouts that can be measured, typically multiple antibody-based markers of phospho-proteins used as proxies of activation[14] or phenotypic readouts such as apoptosis.[15] This can be overcome by using unbiased mass-spectrometry phosphoproteomic data.[16] Furthermore, a similar strategy can be used to identify drug-specific activated pathways from gene expression by connecting drug targets to

TABLE 4.2   List of Resources for the Pipeline

|  | Type | Input Data | Tool Webpage | Publications |
|---|---|---|---|---|
| Macau | Machine learning | Any | https://github.com/ jaak-s/macau | 3, 7 |
| PROGENy | Pathway activity estimation | Transcriptomics | https://saezlab.github. io/progeny/ | 4 |
| DoRothEA | TF activity estimation | Transcriptomics | https://saezlab.github. io/DoRothEA/ | 6 |
| CellNopt | Logic modelling | Perturbation experiments | www.cellnopt.org | 9, 10 |
| Carnival | Logic modelling | Transcriptomics | https://saezlab.github. io/CARNIVAL/ | NA |
| OmniPath | Signaling network database | NA | http://omnipathdb.org/ | 8 |

altered Transcription Factors[17] with the tool CARNIVAL (see Table 4.2). Related tools from other groups can be found elsewhere.[2-4]

On the experimental front, the cancer cell line model confers an unparalleled advantage in terms of ease of study and manipulation. Being able to treat identical entities with hundreds of different perturbations and conditions is a powerful way to understand complex systems. However, some key aspects of cancer drug response are not recapitulated by cell lines. Recently developed model systems allow large-scale screening on more realistic cancer models, e.g. organoids,[18] which take into consideration the organ's 3D structure. Furthermore, direct drug treatment on cancer biopsies[19] (e.g. microfluidics platform[20]) is becoming feasible, allowing study of the tumour microenvironment and the role of the immune system. The approaches described here for cell lines can be easily adapted to these new data modalities.[15]

## Useful Resources

We summarize our resources in Table 4.2.

## References

1. Costello, J.C. et al. A community effort to assess and improve drug sensitivity prediction algorithms. *Nat. Biotechnol.* **32**, 1202–1212 (2014).
2. Ali, M. & Aittokallio, T. Machine learning and feature selection for drug response prediction in precision oncology applications. *Biophys. Rev.* **11**, 31–39 (2018).

3. Zañudo, J.G.T., Steinway, S.N. & Albert, R. Discrete dynamic network modeling of oncogenic signalling: Mechanistic insights for personalized treatment of cancer. *Curr. Opin. Syst. Biol.* **9**, 1–10 (2018).
4. van Hasselt, J.G.C. & Iyengar, R. Systems pharmacology: Defining the interactions of drug combinations. *Annu. Rev. Pharmacol. Toxicol.* **59**, 21–40 (2018).
5. Iorio, F. et al. A landscape of pharmacogenomic interactions in cancer. *Cell* **166**, 740–754 (2016).
6. Schubert, M. et al. Perturbation-response genes reveal signalling footprints in cancer gene expression. *Nat. Commun.* **9**, 20 (2018).
7. Yang, M. et al. Linking drug target and pathway activation for effective therapy using multi-task learning. *Sci. Rep.* **8**, 8322 (2018).
8. Cantini, L. et al. Classification of gene signatures for their information value and functional redundancy. *NPJ Syst. Biol. Appl.* **4**, 2 (2018).
9. Garcia-Alonso, L. et al. Transcription factor activities enhance markers of drug sensitivity in cancer. *Cancer Res.* **78**, 769–780 (2018).
10. Simm, J. et al. Macau: Scalable Bayesian factorization with high-dimensional side information using MCMC. In *2017 IEEE 27th International Workshop on Machine Learning for Signal Processing (MLSP)* 1–6 (2017).
11. Türei, D., Korcsmáros, T. & Saez-Rodriguez, J. OmniPath: Guidelines and gateway for literature-curated signalling pathway resources. *Nat. Methods* **13**, 966–967 (2016).
12. Saez-Rodriguez, J. et al. Discrete logic modelling as a means to link protein signalling networks with functional analysis of mammalian signal transduction. *Mol. Syst. Biol.* **5**, 331 (2009).
13. Terfve, C. et al. CellNOptR: a flexible toolkit to train protein signalling networks to data using multiple logic formalisms. *BMC Syst. Biol.* **6**, 133 (2012).
14. Eduati, F., Doldàn-Martelli, V., Klinger, B. & Cokelaer, T. Drug resistance mechanisms in colorectal cancer dissected with cell type–specific dynamic logic models. *Cancer Res.* **77**, 3364–3375 (2017).
15. Eduati, F., Jaaks, P., Merten, C.A. & Garnett, M.J. Patient-specific logic models of signalling pathways from screenings on cancer biopsies to prioritize personalized combination therapies. *bioRxiv* (2018).
16. Terfve, C.D.A., Wilkes, E.H., Casado, P., Cutillas, P.R. & Saez-Rodriguez, J. Large-scale models of signal propagation in human cells derived from discovery phosphoproteomic data. *Nat. Commun.* **6**, 8033 (2015).
17. Melas, I.N. et al. Identification of drug-specific pathways based on gene expression data: application to drug induced lung injury. *Integr. Biol.* **7**, 904–920 (2015).
18. Drost, J. & Clevers, H. Organoids in cancer research. *Nat. Rev. Cancer* **18**, 407–418 (2018).
19. Letai, A. Functional precision cancer medicine-moving beyond pure genomics. *Nat. Med.* **23**, 1028–1035 (2017).
20. Eduati, F. et al. A microfluidics platform for combinatorial drug screening on cancer biopsies. *Nat. Commun.* **9**, 2434 (2018).

## 4.4 FRAMEWORK FOR HIGH-THROUGHPUT PERSONALIZATION OF LOGICAL MODELS USING MULTI-OMICS DATA

*Jonas Béal, Arnau Montagud, Pauline Traynard,*
*Emmanuel Barillot and Laurence Calzone*

### 4.4.1 Summary

Mathematical models of cancer pathways are built by mining the literature for relevant experimental observations or extracting information from pathway databases to study successive events of tumourigenesis. As a consequence, these models generally do not capture the heterogeneity of tumours and their therapeutic responses. We present here a novel framework, PROFILE, to tailor logical models to particular biological samples such as patient tumours, compare the model simulations to individual clinical data such as survival time, and investigate therapeutic strategies.

Our approach makes use of MaBoSS, a tool based on the Monte-Carlo kinetic algorithm to perform stochastic simulations on logical models resulting in model state probabilities. This semi-quantitative framework allows the integration of mutation data, copy number alterations (CNA) and expression data (transcriptomics or proteomics) into logical models. These personalized models are validated by comparing simulation outputs with expression of patient biomarkers, or clinical data, and then used for patient-specific high-throughput screenings investigating the effects of new mutations or drug combinations. Our approach aims to combine the mechanistic insights of logical modelling with multi-omics data integration to provide patient-relevant models. This work leads to the use of logical modelling for precision medicine and will eventually facilitate the choice of patient-specific drug treatments.

### 4.4.2 Introduction

Mathematical models serve as tools to answer biological questions in a formal way, to detect blind spots and thus better understand a system, to organize information dispersed in different articles into a consensual and compact manner, to identify new hypotheses and to test experimental hypotheses and predict their outcome. In short, a mathematical model can help reason on a problem. Logical models are becoming more popular for exploring cell fate decisions, or particular dysfunctions in biological processes[1] and are particularly appropriate when the question is qualitative, e.g. which genetic mutations can lead to an increase of cell proliferation, which genomic alterations need to be combined to cause resistance to some particular drugs or combinations thereof, etc. Logical models developed specifically in cancer studies are usually used such that asymptotic behaviours, i.e. stable states or limit cycles, are compared to biologically relevant phenotypes. Many software tools exist to study logical models and among them, MaBoSS (Markovian Boolean Stochastic Simulator) is applied to obtain probabilities for each of these asymptotic states using continuous time Markov chain simulations on the logical network[2,3].

In the present work, we underline the importance of including high-throughput analyses in systems-wide modelling in cancer. High-throughput computing, defined as the use of many computing resources over long periods of time to accomplish a computational task, has been extremely useful for Bioinformatics sequence-based analyses enabling its scaling up to multiple genomes. Mathematical modelling in cancer could benefit from the boost in capacities that Bioinformatics profited from.

We hereby present uses of the MaBoSS framework to personalize logical models to a set of patient data, to thoroughly study mutants and optimize drug treatments. Using MaBoSS, a given logical model can be tailored to an omics dataset following our PROFILE (PeRsonalization OF logIcaL modEls) methodology, that can integrate mutations, CNA and expression data, both transcriptomic and proteomic data, if available. Furthermore, all single and double mutants of a given logical model can be automatically simulated to perform perturbation studies on the phenotype probability using our logical modelling pipeline. Finally, MaBoSS can predict the effects of drugs by gradually varying the strength of nodes in a logical model and study their synergies. Taking these three developments into account, researchers can perform patient-specific perturbation analyses and drug treatment designs that are the first step towards precise and personalized medicine.

### 4.4.3 Approach and Application Example

Our framework was built to perform analyses able to retrieve as much information as possible from logical models in the context of cancer-related biological projects. One can either build a new logical model or choose one that corresponds to the subject of interest among the published models in the literature or in repositories[4]. In any case, models should be formalized so that all biological entities (genes, RNA, proteins, etc.) are assigned logical rules that regulate their binary state, either active (1) or inactive (0). In many of these models, environmental determinants are included as inputs and phenotypes as outputs and the questions remain often qualitative, e.g. what combination of gene mutations would enhance a given phenotype or what environmental variable is key in this other phenotype.

These models are usually generic in the sense that they have been designed to be consistent with databases and literature, either summarizing pan-cancer knowledge[5] or focusing on cancer-specific evidences[6], therefore not providing patient-specific insights. Here, we showcase the use of three high-throughput technologies that are built around MaBoSS: Tailoring models to patient and cell-line data, then performing extensive perturbation analyses and studying the system-wide synergistic effects of varying nodes' strengths. To achieve these objectives, our PROFILE methodology generates patient-specific models combining generic logical models and different omics data for patients or cell lines.

The personalization of models is based on different successive steps (Figure 4.4). The first step consists of processing the biological data in order to make them compatible with the logical formalism used by MaBoSS. In this stochastic formalism, the choice of asynchronous updates is made with a Gillespie algorithm, favouring transitions (i.e. node activation or inhibition) with higher transition rates. The stochastic nature of the algorithm makes it relevant to simulate thousands of trajectories, which can then be summarized to obtain an average trajectory where each node has a continuous activation probability between 0 and 1. In this context, it is therefore possible to customize the models using binary data (setting the node value to either 0 or 1 in a simulation), continuous data between 0 and 1 (setting the node value to an average value), or simply continuous and positive constants (as transition rates prioritizing certain reactions with respect to others). Discrete data (such as mutations and CNA) are thus binarized through functional inference: Inactivating (resp. activating) mutations are assumed to correspond to loss (resp. gain) of function mutations and therefore the corresponding nodes of the model are forced to 0 (resp. 1). These inferences are made on the basis of dedicated databases[7] or software[8] and leave many mutations unassigned. Similarly, one can choose to translate the effects of significant copy losses or gains into such binary values.

The processing of continuous data is done in comparison to other patients of the cohort or cell lines in the collection. For each gene, the data distribution pattern is classified as either bimodal, unimodal or zero-inflated. Based on this classification, different normalization methods are proposed (respectively Gaussian mixture models, sigmoid and linear normalization) to convert omics data into values between 0 and 1. The main idea is to preserve distribution patterns, which often carry underlying biological realities (e.g. the bimodal pattern of ERG RNA levels correlate to the reported fusion status of this gene in prostate cancer).

It is possible to personalize logical models with different methods. For instance, defining strict node variants (strict NV) corresponds to setting the selected nodes to a given value throughout the whole simulation, without consideration for the logical rules. This requires binary data that can be obtained from mutations or CNA profiles. Another example considers tuning both initial conditions and transition rates of nodes, thus defining soft node variants (soft NV), which are nodes whose initial node activation value is close to the value of the corresponding omics patient data and whose transition rates promote the maintenance of its state of activity. Since both initial conditions and transitions rates have continuous values between 0 and 1, patient profiles derived from processed RNA/protein levels can thus be used.

Once models have been personalized, their relevance must be verified. The simulations of patient-specific models result in patient-specific probabilities for the phenotypic (output) nodes. The correlation between the probabilities of the model phenotypes and biological biomarkers can then be studied. For instance, one could ask questions such as: 'Is the *Proliferation* simulated score well correlated to the MKI67 levels?'. In general, the correlation of the different phenotypic scores from personalized models can be computed using phenotype-specific RNA signatures using large gene sets (more than 100 genes) from the MSigDB database[9]. Furthermore, clinical studies can also be performed studying the prognostic value of *Proliferation* or *Apoptosis* simulated scores and comparing them to survival data and patient stratification (Figure 4.4)

PROFILE methodology opens the possibility to study more in-depth the heterogeneity of patient behaviour and therapeutic response among a cohort. A first approach consists in simulating new mutations not present in the experimental data in order to better understand the fragility and robustness of the different patient models. High-throughput mutation screening, using additional strict NV, can be performed for each patient

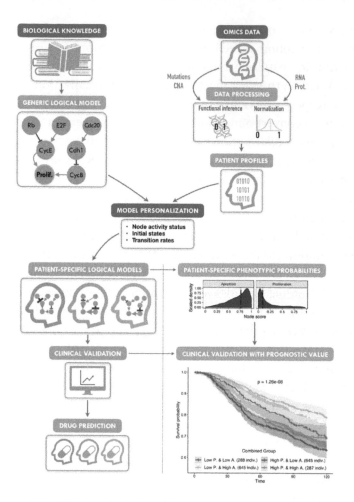

FIGURE 4.4   PROFILE methodology for personalization of logical models (blue boxes) and corresponding results (grey boxes). First, the generic logical model is selected to serve as the starting-point for future personalized models (upper left branch). Then, omics data are gathered (e.g. genome and transcriptome) and processed through functional inference methods (especially for already discrete genome data) or binarization/normalization (for continuous expression data) (upper right branch). The resulting patient profiles are used to perform model personalization, i.e. to adapt the generic model with patient data. Then, clinical relevance of these patient-specific models can be assessed before providing original and personalized therapeutic strategies and drug predictions. As an example (lower right), a generic cancer model[5] has been personalized with mutations as strict NV and RNA as soft NV for 1,865 breast cancer patients from the METABRIC cohort[11]. The survival data of combined groups of high/low Proliferation or Apoptosis scores were calculated based on the phenotypes' median values and their grouping was found to be significant.

with all single or combination of mutations. Our methodology quickly determines the probabilities for all mutant phenotypes, which can then be compared to the probabilities of the *wild type* case to study the effect of these mutations. Combinations of mutations are of particular interest as they can be used to identify if two alterations are, for instance, synergistic (the alteration of the double mutant has more effect on a phenotype than the sum of the single alterations) or synthetic lethal (the double mutant is not viable, while single mutants are viable).

Apart from being able to perform knock-outs, our methodology allows also for knock-down analyses by gradually varying the strength of the nodes, allowing for the simulation of drugs that inhibit the nodes of a logical model[3,10]. As drugs usually have an inhibitory effect on genes, these can be simulated by setting the target node's value to 0 for a complete knock-out effect or to a value between 0 and 1 for a gradual knock-down effect. Combinations of drugs are thus easily modelled with series of simulations where slightly modified node strength variations are applied and this can be upscaled to all the nodes present in a model and with different initial conditions. The results of this set of MaBoSS simulations will be different phenotype probabilities for each one of the knock-downs studied. Again, the probabilities of these knock-downs can be compared to the *wild type* case in order to see the extent of the different phenotype probabilities' shifts.

This fine study of the model nodes' activation helps us identify how a drug can hamper or promote a given phenotype or if the drug action is condition-dependent. Additionally, these analyses can be combined to study the interaction of varying two nodes simultaneously and identifying synergies among them, such as Bliss Independence, or if a phenotype is mostly affected by one of the drugs, but not the other, all this in a patient-specific way using PROFILE-personalized logical models.

### 4.4.4 Discussion and Perspectives

In order to reach its full potential, personalized medicine needs precise mathematical models that are tailored to the data for a given patient. The methodologies presented here are the first steps towards such a reality: The personalization of a logical model to different patient profiles such that their results can be matched to clinical data and then used as a tool for therapeutic investigation. We hereby present three different methodologies that take advantage of the flexibility and potential of MaBoSS to expand the uses of logical models.

Our PROFILE methodology allows the building of precise mathematical models that captures the heterogeneity of patient profiles and their diverse behaviours. These logical models, which are properly specified with patient information, allow researchers to have surrogate mathematical constructs of where to study patients' heterogeneity and different behaviours. The provided code makes the method user-friendly, requiring only a sufficiently large model with signalling pathways of interest that include key genes.

These data-tailored models can be used to study all knock-out or over-expressed mutants for each node and their combinations. Their results are compared to a *wild type* condition to identify which perturbations influence a given phenotype probability. Furthermore, extending our PROFILE methodology for drugs and knock-downs, patient-specific models allow the identification of which drug combinations and their respective levels will have the most effect on a given patient. This analysis is enhanced when the models used are comprehensive and include many signalling pathways and drug-targetable proteins.

These three tools presented here, freely available on GitHub, provide a way to thoroughly personalize models, test mutations and identify druggable points of intervention. They can be of great help to study patient-tailored drug combinations and they enable researchers to test personalized therapeutic strategies *in silico*, paving the way for precision medicine.

## Useful Resources

https://github.com/sysbio-curie/Logical_modelling_pipeline

https://github.com/sysbio-curie/PROFILE

https://github.com/sysbio-curie/drugPROFILE

## References

1. Grieco, L., Calzone, L., Bernard-Pierrot, I., Radvanyi, F., Kahn-Perles, B. & Thieffry, D. Integrative modelling of the influence of MAPK network on cancer cell fate decision. *PLoS Computational Biology* **9**(10), e1003286. (2013).
2. Stoll, G., Viara, E., Barillot, E. & Calzone, L. Continuous time Boolean modelling for biological signalling: Application of Gillespie algorithm. *BMC Systems Biology* **6**(1), 116. (2012).

3. Stoll, G., Caron, B., Viara, E., Dugourd, A., Zinovyev, A., Naldi, A., ... & Calzone, L. MaBoSS 2.0: An environment for stochastic Boolean modelling. *Bioinformatics* **33**(14), 2226–2228. (2017).
4. Chelliah, V., Juty, N., Ajmera, I., Ali, R., Dumousseau, M., Glont, M., ... & Lloret-Villas, A. BioModels: Ten-year anniversary. *Nucleic Acids Research* **43**(D1), D542–D548. (2014).
5. Fumia, H.F. & Martins, M.L. Boolean network model for cancer pathways: Predicting carcinogenesis and targeted therapy outcomes. *PloS One* **8**(7), e69008. (2013).
6. Zañudo, J.G.T., Scaltriti, M. & Albert, R. A network modelling approach to elucidate drug resistance mechanisms and predict combinatorial drug treatments in breast cancer. *Cancer Convergence* **1**(1), 5. (2017).
7. Chakravarty, D., Gao, J., Phillips, S., Kundra, R., Zhang, H., Wang, J., ... & Chang, M.T. OncoKB: A precision oncology knowledge base. *JCO Precision Oncology* **1**, 1 16. (2017).
8. Adzhubei, I., Jordan, D.M. & Sunyaev, S.R. Predicting functional effect of human missense mutations using PolyPhen-2. *Current Protocols in Human Genetics* **76**(1), 7–20. (2013).
9. Liberzon, A., Birger, C., Thorvaldsdóttir, H., Ghandi, M., Mesirov, J.P. & Tamayo, P. The molecular signatures database hallmark gene set collection. *Cell Systems* **1**(6), 417–425. (2015).
10. Flobak, Å., Baudot, A., Remy, E., Thommesen, L., Thieffry, D., Kuiper, M. & Lægreid, A. Discovery of drug synergies in gastric cancer cells predicted by logical modelling. *PLoS Computational Biology* **11**(8), e1004426. (2015).
11. Curtis, C., Shah, S.P., Chin, S.F., Turashvili, G., Rueda, O.M., Dunning, M.J., ... & Gräf, S. The genomic and transcriptomic architecture of 2,000 breast tumors reveals novel subgroups. *Nature* **486**(7403), 346. (2012).

## CHAPTER 5

# Single-Cell Analysis in Cancer

## 5.1 TRACING STEM CELL DIFFERENTIATION WITH SINGLE-CELL RESOLUTION

*Josip Stefan Herman, Dominic Grün*

### 5.1.1 Summary

Single-cell analysis techniques offer us the opportunity to gain unprecedented insights into cellular organs and tissues. The acquisition of transcriptome data for thousands of genes across thousands of individual cells from a tissue of interest makes it possible to study differentiation processes in an unbiased manner. Due to the complexity of these high-dimensional data, computational methods are crucial for the identification of cell types and states and for the inference of relationships between them. Here we give a brief overview of computational lineage tree inference methods using large-scale single-cell RNA-sequencing datasets. We further present as examples StemID,[1] an algorithm for lineage inference and stem cell prediction, and FateID,[2] an algorithm for fate bias inference in progenitor populations.

## 5.1.2 Introduction

Multicellular organisms consist of different cell types that give rise to a multitude of organs and tissues with distinct functions. These cell type-specific functions are acquired during development in a process called cellular differentiation, whereby pluripotent stem cells undergo a sequence of gene expression changes to give rise to all mature cell types. Moreover, in adult organisms tissue-resident stem cells remain crucial for tissue homeostasis in organs with high turnover such as skin, gut or blood, where mature cell types constantly need to be replenished within a few days to maintain organ function.[3] Yet in other organs, such as the liver, stem cells show lower turnover at homeostatic conditions, but can boost their proliferation significantly to regenerate tissue after injury.[4]

With the development of single-cell sequencing techniques in recent years, it is now possible to gain unprecedented insights into the differentiation dynamics from stem cells to mature cell types and resolve cellular heterogeneity of mature cell types, as was shown for the intestine,[5] hippocampus[6] or hematopoietic compartment.[7] Cells from a system of interest are dissociated into single-cell suspensions in most single-cell sequencing protocols and RNA, DNA and/or proteins are barcoded on a single-cell basis using DNA oligonucleotides.[8] Although these single-cell suspension approaches currently offer more options for downstream processing, they come with the downside of erasing spatial information and capturing cell states only at one particular time point. Nevertheless, various computational methods can be used to infer differentiation dynamics and trajectories using pseudo-temporal ordering of these snapshot data, given that enough intermediate states have been sampled during data acquisition. Many of the existing computational methods make use of dimensional reduction approaches such as principal component analysis (PCA), classical multidimensional scaling or diffusion maps to reduce complexity prior to differentiation topology inference. Some of these methods are tree-based while others are graph-based.[9,10] Another group of lineage inference methods, comprising tools such as STEMNET[11] and FateID,[2] uses a probabilistic approach to classify every cell in a dataset according to its probability to differentiate towards a mature cell state without using cluster partition information. In the following section we are going to present the algorithms StemID for lineage inference and stem cell identification and FateID for fate bias inference in progenitor populations.

### 5.1.3 Approach and Application Example

Lineage inference can be a challenging task, especially when multiple differentiation trajectories from a common progenitor pool towards several different mature cell types exist. In this regard the StemID algorithm was developed to infer multiple branching points of cellular differentiation and to predict a potential stem cell phenotype in the dataset. The lineage tree inference of StemID is guided by a previously derived topology of cell types and cell states using a clustering inferred by RaceID3.[2] Medoids of clusters are connected amongst each other, representing potential lineage trajectories and cells are mapped onto the link that best represents its state of differentiation. This is done by projecting the vectors that connect the medoid and the cells in the same cluster, onto the links to all other cluster medoids. In order to find the most likely position of a cell on one of the links, all the link lengths between medoids are normalized to one and the longest projection is chosen. Moreover, to determine whether a link could indeed represent a potential differentiation trajectory and not transcriptional noise, every link is scored according to how densely it is packed. To achieve this, a link score is defined as one minus the maximum difference of the distance between two neighbouring cells on a link after normalizing the link length to one. A score of one would indicate a link that is densely populated with cells with small gaps, whereas a score close to zero would represent cells that are close to the cluster centres of a given link. Apart from lineage tree inference, StemID also aims to predict multipotent cells, namely stem cells, within the data. To this end the concept of entropy is used as a measure of transcriptional uniformity. In general, stem cells tend to express many different transcripts, whereas differentiated cells often express a smaller number of genes at high levels representing their specialized phenotype. This can be modelled using Shannon's entropy where a multipotent progenitor cell with a uniform transcriptome would show high entropy values, whereas a differentiated cell and its specialized transcriptome would have low entropy values. Finally, StemID uses the product of both the number of outgoing branching links and the entropy values of a given cell type to calculate a score reflecting the likelihood of a given cell type to be a stem cell. Despite being useful in inferring lineage trajectories and predicting stem cells, StemID is dependent on a pre-inferred topology.

However, newly developed methods, such as the FateID algorithm, are circumventing this by using a probabilistic approach. FateID was developed to infer fate biases towards different cell lineages. Fate bias can be understood as the probability of a given cell to follow a differentiation path towards a specific mature cell state. In an iterative process the algorithm classifies the fate biases towards a number of given mature cell states. For this purpose, an expression matrix of all genes and cells alongside a partitioning of these mature (target) states is provided to the algorithm. FateID will start by using these target states as a training set for an iterative random forests classification[12] in order to learn their defining features. At each step of the iteration, a fixed number of cells in the neighbourhood of every target state with the smallest median distance to this state will be selected. These newly selected cells will serve as the test set and will be classified based on the training set using random forests. Finally, cells with significantly more votes towards one of the target states than all other target states will be included in the training set of the next iteration. Moreover, at each iteration the training set will only comprise a set of cells closest to the test set. In this regard, FateID also allows removing more mature cells as newly classified cells enter the training set. At each iteration, this dynamic training set progresses backwards in differentiation time to classify less mature cell states based on their more mature neighbouring cell states and makes it possible to track the importance of particular genes for each cell state along a pseudo-temporal trajectory that might be lost when classifying all cells solely based on the mature states. FateID has been successfully applied in recent studies.[13]

In this respect, one of the best-studied differentiation processes in mammals is hematopoiesis where a multipotent hematopoietic stem cell (HSC) generates multipotent progenitors (MPP) that further differentiate to mature cell types such as erythrocytes, megakaryocytes, granulocytes, monocytes, dendritic cells, B cells, T cells or innate lymphoid cells. StemID and FateID were used on a recent dataset from hematopoieitc progenitors[2] (Figure 5.1). Conclusively, StemID was able to infer the hematopoieitc lineage tree by connecting different stages of progenitor cells to mature cell types. However, it is noteworthy that some of the less significant links (in green colour) will likely represent artefacts of the analysis, that can be attributed, for example, to transcriptional similarities of mature cell types with similar functions. On the other hand, FateID inferred overlapping domains of fate bias towards the distinct mature cell types within the multipotent progenitor compartment, which could be validated through in vitro experiments.[2]

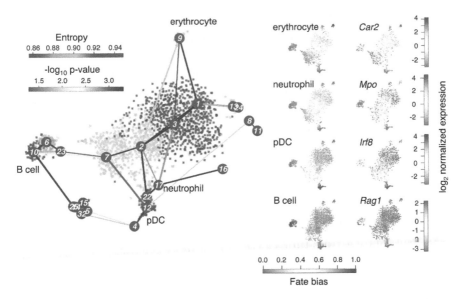

FIGURE 5.1    The left panel shows a t-SNE map representation of bone marrow hematopoietic single-cell transcriptome data and highlights clusters inferred by RaceID3 in different colours. Every dot represents one cell. The lineage tree that was inferred by StemID is overlaid and depicts all links with a link score greater than 0.4. Node numbers indicate cluster medoids. Entropy levels and negative $\log_{10}$ p-values for the inter-cluster links are represented using gradient colour scales. The right panel shows the fate biases towards target clusters using colour gradients from blue (low fate bias) to red (high fate bias). Fate biases and $\log_2$-transformed expression of known marker genes are shown for target clusters 9 (erythrocytes), 17 (neutrophils), 12 (plasmacytoid dendritic cells, pDC) and 10 (B cells).

## 5.1.4 Discussion and Perspectives

The rapid progress of single-cell sequencing techniques has led to continuously growing interest in investigating biological systems on a single-cell basis and has given rise to international efforts to map all human cell types to create a human cell atlas.[14] In a similar fashion, the development of single-cell analysis techniques accelerated in order to harness the tremendous potential of single-cell data using different clustering, batch-effect removal or pseudo-temporal ordering methods to mention a few. Increasing numbers of lineage inference algorithms have been developed and there are attempts on standardizing benchmarking criteria.[15] Especially with the constantly increasing numbers of cells and cell types being sequenced, lineage inference methods need to run computationally efficient and need to distinguish well between recurring expression of genes such as cell cycle genes or lymphocyte recombination genes that might confound lineage inference. Furthermore, recent approaches like approximate graph abstraction[16] attempting to reconcile clustering and trajectory inference and trajectory inference combined with force-directed layouts[17] could hold great potential to better understand single-cell snapshot data. Moreover, techniques such as the measurement of RNA velocity to derive instantaneous rates of change in RNA transcription by comparing the ratios of spliced versus unspliced RNA can be very valuable for lineage inference methods by adding a layer of transcriptional directionality to the data.[18] Despite the importance of computational methods in gaining insights into pseudo-temporal lineage hierarchies, it is also important to acquire actual time-resolved data, e.g. from time course experiments. Alternatively, innovative CRISPR/Cas9-based methods allow the introduction of unique genomic alterations in a cell at discrete time points *in vivo* and thereby marking all of its progeny revealing ancestral relations between cell types.[19,20] Future developments of these techniques might even allow the recording of time in a continuous fashion and replace pseudo-time with real-time. We are only at the beginning of leveraging the full potential of single-cell analysis techniques and will make many more exciting discoveries in biological research and medicine.

## Useful Resources

RaceID3 and StemID2

https://cran.r-project.org/web/packages/RaceID/vignettes/RaceID.html

FateID

https://cran.r-project.org/web/packages/FateID/vignettes/FateID.html

## References

1. Grün, D. et al. De novo prediction of stem cell identity using single-cell transcriptome data. *Cell Stem Cell* **19**, 266–277 (2016).
2. Herman, J.S., Sagar & Grün, D. FateID infers cell fate bias in multipotent progenitors from single-cell RNA-seq data. *Nat. Methods* **15**, 379–386 (2018).
3. Pellettieri, J. & Alvarado, A.S. Cell turnover and adult tissue homeostasis: From humans to planarians. *Annu. Rev. Genet.* **41**, 83–105 (2007).
4. Stanger, B.Z. Cellular homeostasis and repair in the mammalian liver. *Annu. Rev. Physiol.* **77**, 179–200 (2015).

5. Grün, D. et al. Single-cell messenger RNA sequencing reveals rare intestinal cell types. *Nature* **525**, 251–255 (2015).

6. Hochgerner, H., Zeisel, A., Lönnerberg, P. & Linnarsson, S. Conserved properties of dentate gyrus neurogenesis across postnatal development revealed by single-cell RNA sequencing. *Nat. Neurosci.* **21**, 290–299 (2018).

7. Paul, F. et al. Transcriptional heterogeneity and lineage commitment in myeloid progenitors. *Cell* **163**, 1663–1677 (2015).

8. Grün, D. & van Oudenaarden, A. Design and analysis of single-cell sequencing experiments. *Cell* **163**, 799–810 (2015).

9. Grün, D. Revealing routes of cellular differentiation by single-cell RNA-seq. *Curr. Opin. Syst. Biol.* **11**, 9–17 (2018).

10. Herring, C.A., Chen, B., McKinley, E.T. & Lau, K.S. Single-cell computational strategies for lineage reconstruction in tissue systems. *Cell. Mol. Gastroenterol. Hepatol.* **5**, 539–548 (2018).

11. Velten, L. et al. Human haematopoietic stem cell lineage commitment is a continuous process. *Nat. Cell Biol.* **19**, 271–281 (2017).

12. Breiman, L. Random forests. *Mach. Learn.* **45**, 5–32 (2001).

13. Weinreb, C., Rodriguez-Fraticelli, A.E., Camargo, F.D. & Klein, A.M. Lineage tracing on transcriptional landscapes links state to fate during differentiation. *bioRxiv* (2018). doi:10.1101/467886.

14. Regev, A. et al. The human cell atlas. *eLife* **6**, (2017).

15. Saelens, W., Cannoodt, R., Todorov, H. & Saeys, Y. A comparison of single-cell trajectory inference methods: towards more accurate and robust tools. *bioRxiv* (2018). doi:10.1101/276907.

16. Wolf, F.A. et al. Graph abstraction reconciles clustering with trajectory inference through a topology preserving map of single cells. *BioExiv* (2018). doi:10.1101/208819.

17. Weinreb, C., Wolock, S. & Klein, A.M. SPRING: A kinetic interface for visualizing high dimensional single-cell expression data. *Bioinformatics* **34**, 1246–1248 (2018).

18. La Manno, G. et al. RNA velocity of single cells. *Nature* **560**, 494–498 (2018).

19. Spanjaard, B. et al. Simultaneous lineage tracing and cell-type identification using CRISPR–Cas9-induced genetic scars. *Nat. Biotechnol.* **36**, 469–473 (2018).

20. Kalhor, R. et al. Developmental barcoding of whole mouse via homing CRISPR. *Science* **361**, t9804 (2018).

## 5.2 PHYLOGENY-GUIDED SINGLE-CELL MUTATION CALLING

Jochen Singer, Katharina Jahn, Jack Kuipers, Niko Beerenwinkel

### 5.2.1 Summary

The development of single-cell sequencing technologies offers unprecedented resolution in tumour phylogeny reconstruction. However, single-cell sequencing suffers from elevated error rates, non-uniform coverage and allelic dropout, such that traditional approaches for mutation calling, designed for bulk sequencing data, are ill-suited. Here we describe SCIΦ, a mutation calling approach tailored to single-cell sequencing data, which leverages information between cells. SCIΦ integrates mutation calling with inferring the underlying phylogenetic cell lineage tree in a fully Bayesian modelling approach. It robustly accounts for single-cell sequencing artefacts and reliably calls mutations even in low-coverage regions. Using data from a breast cancer tumour, we demonstrate the applicability of SCIΦ to single-cell sequencing data and its potential to gain insights into tumour evolution and heterogeneity.

### 5.2.2 Introduction

Cancer is a severe disease caused by mutations in the genome accumulated in individual cells. Some of these mutations can provide a proliferation advantage to the cell compared to its neighbours, allowing the clone formed by the cell and its progeny to expand and give rise to a tumour.[1] Over the course of time, tumour cells can acquire additional mutations that lead to the formation of subclones.[1] This evolutionary process results in genetic diversity within a single tumour, referred to as intra-tumour heterogeneity, which plays a central role in the failure of targeted cancer therapies.[2] For example, new resistant mutations may emerge, or subclones that were suppressed before treatment may start to expand. Even for monoclonal tumours, it has been shown that the temporal order of mutation occurrence can be informative for drug therapy.[3] Therefore, a better understanding of tumourigenesis is key for more efficient and effective personalized cancer treatment.

Until recently, studies inferring tumour phylogenies used bulk sequencing approaches. Bulk sequencing retrieves the genetic information from a mixture of thousands to millions of cells simultaneously. In doing so, the knowledge of which sequencing read came from which cell is lost. Bulk sequencing approaches therefore need to deconvolve the information in order to infer the unknown subclones, their frequencies and underlying phylogeny, such that their resolution is typically very limited.[4]

In contrast, single-cell sequencing technologies offer a direct mutation-to-cell assignment, which provides new opportunities for tumour phylogeny reconstruction.[5,6] However, due to the limited genomic material in each cell, single-cell sequencing requires heavy amplification, which results in elevated error rates, uneven coverage, drop-outs and missing information.[7] To address these challenges, we developed the SCITE family of statistical models and computational tools, including SCITE[8] to reconstruct tumour phylogenies, $\infty$SCITE[9] to allow for recurrent mutations, and SCI$\Phi$[10] to infer the tumour phylogeny and call mutations simultaneously.

Here we focus on mutation calling by sharing the nucleotide information across cells to call mutations and infer the sample phylogeny jointly. By sharing information across the phylogeny, SCI$\Phi$ is able to augment the evidence for a mutation detected in a single cell with information on the same nucleotide change observed in closely related cells.

### 5.2.3 Approach and Application Example

In contrast to other single-cell mutation callers, SCI$\Phi$ not only reports the mutation status of the cells, but also provides information on the tumour phylogeny, which allows insights into tumour heterogeneity and evolution. SCI$\Phi$ is based on a probabilistic model that allows assessing the joint posterior distribution of mutation calls and phylogenetic tree given observed single-cell sequencing data. The posterior probabilities of the mutation-to-cell assignments quantify the confidence in the mutation calls and can be used in downstream analyses to take this uncertainty into account.

Because SCI$\Phi$ assumes independence between loci and does not require any phasing information with known or germline SNPs, it is not only applicable to whole-genome sequencing data, but can also be used with exome or panel sequencing data.

SCI$\Phi$ is easy to use and accepts the sequencing information in mpileup format. The default call to SCI$\Phi$ is: sciphi --in cellSpec.txt -o result cells. mpileup, which lets SCI$\Phi$ process the sequencing information provided in cells.mpileup using the cell specifications provided in cellSpec.txt. The file cellSpec.txt contains the cell name in the first column and cell type in the second. The cell type can be BN (bulk normal/control), CN (cell normal/control) or CT (cell tumour). In a two-step procedure, SCI$\Phi$ first identifies candidate loci and then samples phylogenetic tree structures and mutation-to-cell assignments. After completion, SCIΦ will generate result.gv, result.probs and result.vcf. The file result.gv contains the structure of the maximum a posteriori (MAP) tree, result.probs contains the posterior probabilities of the mutation to cell assignment, and result.vcf provides the mutation assignment in variant calling format. Several additional parameters may be specified, for example, to record samples from the full posterior distribution.

The mpileup file can be generated using SAMtools[11] allowing the user to apply quality control filters. For example, for the mpileup generation with SAMtools, we advise setting filters on the base (-Q 30) and mapping quality (-q 40) to obtain high-quality mapping information.

Because in typical tumour analyses the focus is on mutations with a functional impact, the user may provide a normal bulk sequencing control sample to filter out germline mutations during the candidate loci identification phase. In addition, SCIΦ offers the possibility for the user to provide additional loci to ignore, for example, positions of common SNPs, to further tailor the analysis to tumour-specific mutations.

An important feature of SCIΦ is its DNA amplification model. Due to the limited amount of DNA in each cell, the DNA is amplified, often using multiple displacement amplification (MDA).[7] MDA tends to result in very uneven coverage profiles and, more importantly, also in very uneven allele fractions across cells per locus. The MDA process is modelled as a Polya urn, in which an allele is drawn randomly from a cell and replaced together with a copy of itself. This process is repeated until a certain coverage is reached. The Polya urn process results in a beta-binomial distribution, which we employ in SCIΦ to represent the nucleotide counts of single cells. The beta-binomial distribution is also frequently used in bulk mutation calling approaches (e.g. Gerstung[12]) and by learning its parameters from the observed data, in addition to MDA, SCIΦ is also applicable to other amplification methods, e.g. MALBAC or pure PCR based methods.

Similar to bulk sequencing approaches, sequencing and amplification errors may lead to false positive mutation calls. However, in comparison to bulk sequencing, single-cell amplification errors are more common and often reach far greater allele fractions, such that they represent a limiting factor for single-cell genomics. SCIΦ minimizes false positive calls already during candidate loci identification by only considering loci with mutation evidence in at least two cells. Additional false positive mutation calls may result from mapping artefacts or in regions with elevated sequencing error rates. In contrast to other approaches, SCIΦ filters them explicitly by applying a likelihood ratio test that discards mutations that are present in low frequencies across all cells. While low frequencies in some cells are expected, it is unlikely to observe only low-variant allele frequencies for a real mutation.

A major problem when using traditional variant callers on single-cell data is their lack of a model for drop-out events. Drop-out events describe the amplification of only one of the two alleles and are frequently observed at rates of 10 to 50%.[7] Therefore, we learn the drop-out rate of the experiment and integrate it into SCIΦ's statistical model to approximate the single-cell-specific nucleotide distribution.

After candidate loci identification, SCIΦ infers the phylogenetic structure of the sample. To sample from the joint posterior distribution of mutation calls and phylogenetic trees, SCIΦ employs Markov Chain Monte Carlo (MCMC). During MCMC, an initial tree is repeatedly modified and scored to best represent the genomic data of the sample. The starting tree can be provided by the user, for example from a previous analysis, or is chosen randomly. In addition to the phylogenetic tree, during the MCMC scheme, all model parameters are inferred, including those of the beta-binomial distributions and the drop-out rate.

To demonstrate the use of SCIΦ for single-cell analysis, we applied it to a triple-negative (ER-/PR/Her2-) breast cancer data set consisting of 16 tumour cells and a matched bulk control sample.[13] The tumour cells were sorted according to ploidy, and the exomes of eight hypodiploid (h) and eight aneuploid (a) cells were sequenced. Figure 5.2 summarizes the posterior distribution displaying the MAP tree with the average number of mutations assigned to the nodes on the left-hand side and the posterior probabilities of the mutation calls on the right-hand side.

The MAP tree shows the evolutionary history of the sample. More than 800 mutations are assigned to the root node and are therefore predicted to be present in all samples. The large number of clonal mutations indicates the accumulation of mutations in successive clonal expansions. The tree also shows that more than 300 mutations differentiate cell *h1* from the other cells. However, from the posterior sample we learn that a considerable fraction of mutations does not show a clear signal whether the mutation is present in *h1* or not. The reason for this uncertainty is the high fraction of missing information for the corresponding loci for that cell.

In addition, the tree reveals the emergence of four very distinct subclones with many mutations private to a particular clade. For example, 206 mutations are private to the clade of *a1*, *a4* and *a6*, distinguishing these

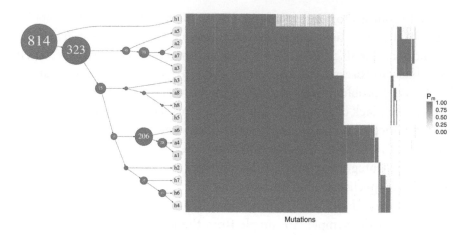

FIGURE 5.2   Phylogenetic tree and posterior mutation probabilities for 16 cells of a triple negative breast cancer whole-exome single-cell sequencing data set. On the left, the maximum a posteriori phylogenetic tree with the average number of mutations attached to the nodes is shown. The node size corresponds to the number of assigned mutations and the leaves are coloured according to ploidy. The right side shows the posterior probabilities of mutation calls.

cells from the clade of *h2*, *h4*, *h6* and *h7*. It is interesting to observe that *a8* clusters together with *h3*, *h5* and *h8*, even though they belong to different ploidy classes. However, as can be observed in Figure 5.2 *a8* shares several mutations with *h3*, *h5* and *h8*, while not sharing any mutations private to the clusters consisting of *a2*, *a3*, *a5* and *a7*, and the cluster of *a1*, *a4* and *a6*. The four main subclones themselves exhibit heterogeneity across their constituent cells, indicating the complexity of the tumour composition, though the snapshot of 16 cells may not be a fully representative sample of the tumour, which could exhibit even greater heterogeneity.

### 5.2.4  Discussion and Perspectives

Single-cell sequencing offers unprecedented resolution for assessing tumour heterogeneity and evolution. However, single-cell sequencing data have elevated noise levels, such that tailored analysis tools are required. SCIΦ jointly calls mutations and infers the underlying sample phylogeny. In doing so, information across cells can be used to improve mutation calls. SCIΦ makes the infinite sites assumption, which states that an acquired mutation is not lost in subsequent cell divisions and that a mutation is not acquired independently more than once. However, it has been shown that this assumption can be violated in tumour samples.[9] Even though SCIΦ shows only a moderate decrease in performance in the presence of some violations of the infinite sites assumption, future versions should weaken, or ideally drop, this assumption. SCIΦ shares genetic information across cells for a particular locus, but not across loci. Phasing information across loci for a particular cell, for instance by using long-read technologies, could be used to more reliably distinguish between mutations and sequencing noise from the amplification process or mapping artefacts. Another desirable extension is the inclusion of copy number events. Due to the non-uniform coverage, copy number events are very challenging to identify in single-cell data. However, if the copy number states were known one could adjust the parameters of the nucleotide models accordingly and retrieve even more reliable mutations calls. Similar arguments hold for the inclusion of structural variants, such that the long-term goal is to include the different types of genomic variants simultaneously and retrieve a complete history of the genomic alterations of the tumour.

## Useful Resources

SCIΦ: https://github.com/cbg-ethz/SCIPhI, SCITE: https://github.com/cbg-ethz/SCITE, ∞SCITE: https://github.com/cbg-ethz/infSCITE

## References

1. Merlo, L.M.F., Pepper, J.W., Reid, B.J. & Maley, C.C. Cancer as an evolutionary and ecological process. *Nat. Rev. Cancer* **6**, 924–935 (2006).
2. Greaves, M. & Maley, C.C. Clonal evolution in cancer. *Nature* **481**, 306–313 (2012).
3. Ortmann, C.A. et al. Effect of mutation order on myeloproliferative neoplasms. *N. Engl. J. Med.* **372**, 601–612 (2015).
4. Van Loo, P. & Voet, T. Single cell analysis of cancer genomes. *Curr. Opin. Genet. Dev.* **24**, 82–91 (2014).
5. Kuipers, J., Jahn, K. & Beerenwinkel, N. Advances in understanding tumour evolution through single-cell sequencing. *Biochim. Biophys. Acta Rev. Cancer* **1867**, 127–138 (2017).
6. Davis, A. & Navin, N.E. Computing tumor trees from single cells. *Genome Biol.* **17**, 113 (2016).
7. Wang, Y. & Navin, N.E. Advances and applications of single-cell sequencing technologies. *Mol. Cell* **58**, 598–609 (2015).
8. Jahn, K., Kuipers, J. & Beerenwinkel, N. Tree inference for single-cell data. *Genome Biol.* **17**, 86 (2016).
9. Kuipers, J., Jahn, K., Raphael, B.J. & Beerenwinkel, N. Single-cell sequencing data reveal widespread recurrence and loss of mutational hits in the life histories of tumors. *Genome Res.* **27**, 1885–1894 (2017).

10. Singer, J., Kuipers, J., Jahn, K. & Beerenwinkel, N. Single-cell mutation identification via phylogenetic inference. *Nat. Commun.* **9**, 5144 (2018).
11. Li, H. et al. The sequence alignment/map format and SAMtools. *Bioinformatics* **25**, 2078–2079 (2009).
12. Gerstung, M. et al. Reliable detection of subclonal single-nucleotide variants in tumour cell populations. *Nat. Commun.* **3**, 811 (2012).
13. Wang, Y. et al. Clonal evolution in breast cancer revealed by single nucleus genome sequencing. *Nature* **512**, 155–160 (2014).

# Patient Stratification and Treatment Response Prediction

## 6.1 INTEGRATIVE NETWORK-BASED ANALYSIS FOR SUBTYPING AND CANCER DRIVER IDENTIFICATION

*Lieven P.C, Verbeke, Maarten Larmuseau, Louise de Schaetzen van Brienen and Kathleen Marchal*

### 6.1.1 Summary

With the decreasing sequencing cost, cohorts of tumour samples are increasingly being subjected to (epi-) genomic profiling that is often complemented with additional functional data like clinical data or gene expression levels. A plethora of methods have been developed to analyze these cohort-derived multi-omics data in order to identify subtypes, driver genes and driver pathways. A promising class of approaches for analyzing cancer systems genetics are *network-based methods*. These methods rely on a prior knowledge in the form of interaction networks and use an explicit network model to drive their analysis. The explicit network model offers the advantage of *providing an intuitive way of integrating* the systems genetics data. Analyzing all data simultaneously allows to fully exploit data complementarity in order to identify subtypes and their causal drivers in one comprehensive analysis. In addition, by searching for recurrently disturbed (containing a driver mutation) network neighbourhoods rather than genes, network-based methods *are ideally suited to also identify rare driver events.* The recovered recurrently mutated network neighbourhoods are proxies for the driver pathways and provide *insight into the mode of action* of the cancer-related phenotype.

## 6.1.2 Introduction

In contrast to gene-centric driver identification methods,[1] network or pathway-based methods do not assume that the same driver gene should be hit in all samples, but rather that the same driver pathways are affected, potentially through different genes in different samples. By searching for recurrently mutated pathways rather than genes, the power of identifying a driver signal increases and rare drivers can indirectly be identified.[2,3] In contrast to pathway-based methods, network-based methods do not require an up-front fixed definition of functional groups of genes, but rather search for network neighbourhoods that are enriched in driver events.[2]

Network methods differ in their underlying algorithms and the type of questions that can be answered. Propagation-based methods are a popular strategy to identify subgraphs enriched in driver events.[4] Typically, propagating a gene-centric signal (e.g. mutational burden or a gene-specific driver score) over the network model results in a data transformation that can be used to prioritize drivers (NBDI,[5,6] NetICS[7]), to find subgraphs on the interaction network enriched for gene-specific properties (HotNet2,[8] NBDI[5]) or for subtyping (Hofree,[9] NBDI,[5] Le van[10]). Significance area detection methods (SASMs)[11] rely on a greedy strategy to identify high scoring subnetworks, consisting of subsets of genes connected on the interaction network. Although the subnetwork score can be a simple aggregate of gene-specific scores (e.g. mutational burden in NetSig[12]), SASMs are ideally used with a score that reflects an intrinsic property of the gene set composing the subnetwork, such as mutual exclusivity (e.g. Memo,[13] Mutex,[14] MEMcover,[15] COVEX,[16] SSA.ME[17]). A last class of methods relies on probabilistic graph models (Phenetic,[18] Paradigm[19,20]). Because of their expressiveness, these models allow for maximally exploiting all information present in the prior interaction network (including directionality and edge weights).

Next to differing algorithmically, network-based methods also differ in the network models they use. Most methods use a model in which genes are nodes and edges represent the relations between the genes, derived from prior interaction information. These simple network models allow for analyzing a single data source (e.g. mutational information) while searching for recurrently affected subnetworks, but do not exploit the expressiveness of network models to integrate multiple datasets. Integrating data allows for leveraging data complementarity and provides more mechanistic insights in the identified drivers, pathways and subtypes, yet few network-based methods currently exist that rely on an integrative network model.[5–7,21]

## 6.1.3 Approach and Application Example

To illustrate the potential of an integrative network-based method, we present MUNDIS, a MUlti-purpose Network-based Data Integration Strategy. By using an integrative network model, MUNDIS allows combining multi-omics data to derive subtypes together with their causal driver genes and pathways. MUNDIS (an extension of NBDI[5]) algorithmically is a propagation method. Its integrative network model is unique in the way different complementary data sources are casted, together with the prior interaction network, into a single integrative model. The connectivity in this network model is leveraged to subtype and prioritize drivers and driver pathways (Figure 6.1A). In contrast to simple network models, the MUNDIS model also contains, next to nodes that represent genes, nodes that represent the individual samples and the aberrant status of the genes in the samples. Including the gene status provides an intuitive way of maximally exploiting different types of data. A gene can be aberrant in many different ways: By being aberrantly expressed (EXP), containing a somatic mutation (MUT), being differentially methylated (MET) or by having an abnormal copy number (CNV). The status of a gene is assumed to be binary (with 1 an aberrant status and 0 a normal status). A gene node is connected to a sample node through its corresponding status nodes (one for each data type included in the analysis). Status nodes are connected to the respective sample nodes in which the represented aberrant status was observed. The information on the a priori interaction network is incorporated through the edges that connect the gene nodes.

The integrative MUNDIS network model summarizes all available data and inherently contains, through its topology, the information needed for multiple applications. *Subtyping* requires grouping samples (patients) based on their connectivity (also called similarity). Sample–sample connectivity is estimated using a propagation-based strategy. The better connected two samples are, the more similar they will be. Samples will be more connected if they share the same affected gene status (e.g. they share a mutated gene, or multiple differentially expressed genes) or if they have an abnormal gene status for different genes that are closely connected in the interaction network (i.e. a recurrently disturbed subnetwork as a

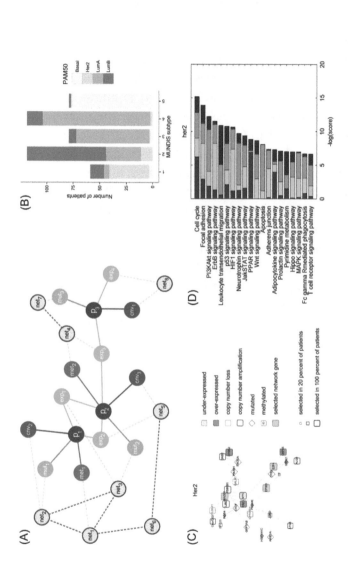

FIGURE 6.1 The MUNDIS network-based method for subtyping, driver gene identification and pathway relevance ranking. Presented results were obtained by applying MUNDIS on a TCGA breast cancer dataset. (A) Integrative network model of the network-based method MUNDIS. Nodes represent genes, samples or an aberrant status of a gene. Edges between gene nodes are derived from the KEGG interaction network, complemented with ENCODE TF-target interactions. Edges between sample and status nodes and between status and gene nodes respectively are derived from binarized methylomics, transcriptomics and genomics (mutation and copy number) data. The connectivity in the network model is exploited using a propagation method to derive scores between any pair of nodes. (B) Scores between sample nodes reflect sample–sample similarities and are used for subtyping. (C) Scores between samples and status or gene nodes respectively reflect the relevance of a gene (gene node) or an aberration (status node) to a sample. These scores are used to prioritize drivers (genes with somatic mutations, CNVs or aberrant methylation) and important differentially expressed genes. Mapping the selected driver genes to the network prior results in connected components that proxy driver pathways. (D) Testing to what extent the average scores of genes that belong to a predefined pathway are higher than what can be expected by chance allows ranking pathways that are relevant to the identified subtypes. The probabilistic score assigned to each pathway can be decomposed into four components. The contribution of each component to the total score is indicated in a different colour bar: mRNA expression (dark blue), copy number (light blue), mutation (pink) and methylation (dark red).

proxy of a driver pathway). A propagation derived score (here derived by applying a Laplacian Exponential Diffusion graph node kernel) estimates the similarity between sample nodes by summarizing their connectivity in the network in a single value. Simply clustering the samples based on these pairwise scores between sample nodes results in integrative subtyping (Figure 6.1B). *Drivers* can be identified in an analogous way by using the connectivity scores, obtained after network propagation, between the sample nodes and respectively the status and gene nodes. The higher the scores, the more relevant the aberrant status and/or gene is for the sample. Genes or status nodes that are relevant to multiple samples of the same subtype (corresponding to a high mean score over the samples of the subtype) are subsequently prioritized as potential drivers. By calculating an average statistic for the different status nodes of the same gene separately, we can assess the extent to which each aberration type (MUT, EXP, CNV, MET) contributes to the samples in the subtype. Note that although the status nodes of specific genes might show a high relevance for certain samples, this does not necessarily imply that the genes are actually in an aberrant status in all of the samples to which they appear relevant. A gene can be relevant to a sample because that sample carries an aberration in a different gene that is closely connected in the interaction network to the gene of interest. Hence, all genes that belong to a recurrently disturbed network neighbourhood will be relevant to all samples that carry at least one aberration in that network neighbourhood. This becomes more apparent when the selected driver genes (genes with an aberrant methylation, copy number or somatic mutation status) are mapped on the interaction network prior (Figure 6.1C). Indeed, the drivers identified for each subtype constitute *specific connected components, which can be considered as subtype-specific driver pathways*. The same network-propagation derived scores can also be used for *pathway importance ranking*, i.e. to identify which pathways – containing active or inactive genes, mutated genes, genes with altered copy number or hyper- or hypo-methylated genes – are more relevant for a group of patients than others. The pathways of interest should be available here as predefined sets of genes. Pathways are ranked by averaging, per data type, the raw connectivity scores between all the genes in the pathway and the patients in the subtype. These averaged scores are converted to (pseudo) p-values using a permutation strategy (for details, see Verbeke et al.[5]).

The results shown in Figure 6.1 were obtained by applying MUNDIS to a subset of the TCGA breast cancer dataset. Data preprocessing, binarization and filtering were done as previously described[5]. The propagation method contains a single quantitative parameter, i.e. the kernel parameter (see[22] for details), which controls how large a neighbourhood will be taken into account by the diffusion kernel. The smaller the parameter, the less important neighbouring genes in the global network will be. For the subtyping task a value of 0.0005 was used; for all other applications we used a value of 0.01. Application of MUNDIS to a breast cancer patient cohort revealed subtypes that were very similar to the PAM50 molecular subtypes (Figure 6.1C), indicating the biological relevance of the subtypes. The per-subtype driver gene prioritization (Table 6.1) recapitulates to a large extent what is known on the molecular biology of breast cancer tumours (see e.g. the Cancer Genome Atlas Network,[23] where the same data were used in combination with different methods). For instance, the *HER2* subtype (case shown here for illustrative purposes, Figure 6.1C, Table 6.1) is determined by the amplification of *ERBB2*, and a combination of mutations in *TP53* and the *PIK3CA* and *PIK3R* kinases. All selected methylated genes were hyper-methylated. The number of methylated genes increases from Basal over Her2 to Luminal A and Luminal B. *LAMA1* (an ECM-receptor gene) is a notable exception on this with a lower methylation frequency for the Luminal subtype compared to the Her2 and Luminal B subtypes. The TCGA breast cancer study[23] identified strong methylation of genes in the WNT-signalling pathway, involved in cell differentiation and proliferation. Of that pathway, for both luminal subtypes, we find *SFRP1*, a WNT antagonist whose methylation condition can be related to patient survival.[24] Additionally, the cyclin *CCND2*, a gene with known oncogenic properties[25] scored high for all four subtypes. The pathway importance ranking obtained by MUNDIS on these breast cancer data (Figure 6.1D) largely confirmed the PARADIGM-based pathway ranking[20] with the top 15 pathways mentioned in the PARADIGM study all mapping to pathways present in the MUNDIS rankings.

This case study shows how an integrative network model intuitively combines the different systems genetics data to simultaneously delineate meaningful subtypes and explain their molecular origin by identifying the relevant drivers and driver pathways.

## 6.1.4 Discussion and Perspectives

Genetic information of cohorts of 1000s of tumour samples is increasingly being complemented with transcriptomic and epigenetic profiling. When coping with such systems genetics data, most studies rely on a sequential analysis in which different aspects of tumour biology and

TABLE 6.1  Driver Gene Prioritization for the Basal-like, Her2, Luminal A and Luminal B Breast Cancer Subtypes

| Copy Number | | | | | Mutation | | | | | Methylation | | | | |
|---|---|---|---|---|---|---|---|---|---|---|---|---|---|---|
| Gene | Basal | Her2 | Luminal A | Luminal B | Gene | Basal | Her2 | Luminal A | Luminal B | Gene | Basal | Her2 | Luminal A | Luminal B |
| AIM1 | 0.025 (14) | 0.113 (7) | | | AKT1* | | | 0.033 (11) | 0.036 (12) | ABO | | 0.774 (9) | | |
| ALG8 | | 0.094 (14) | 0.048 (15) | 0.143 (9) | ALOXE3 | | | | | ADHFE1 | 0.395 (6) | | | |
| AP3M2 | | | 0.048 (13) | 0.125 (13) | ANAPC5 | 0.037 (8) | | | | AKR1B1 | | 0.792 (4) | 0.684 (2) | 0.830 (3) |
| AQP11 | | 0.094 (15) | 0.053 (10) | 0.134 (10) | ARHGEF1 | 0.049 (5) | | | | CCND2* | 0.321 (7) | 0.604 (14) | 0.651 (3) | 0.786 (5) |
| CCDC77 | 0.049 (7) | | | | ARID1A* | | 0.057 (7) | | | CDX2* | | | | 0.804 (10) |
| CCND1* | | 0.170 (5) | 0.110 (1) | 0.321 (1) | ARMCX1 | | 0.038 (14) | | | CLDN11 | | 0.755 (7) | | |
| CHCHD2 | 0.037 (9) | | | | ATN1 | | | 0.029 (13) | | COL23A1 | | | 0.541 (9) | 0.661 (14) |
| CLNS1A | | | 0.053 (11) | 0.143 (6) | ATP1A4 | | 0.094 (3) | | | CYP24A1 | | 0.792 (15) | | |
| ENSA | 0.074 (2) | | | | BRCA1* | 0.062 (2) | | | | DPYS | 0.790 (1) | 0.698 (6) | 0.708 (5) | 0.839 (4) |
| ERBB2* | | 0.679 (1) | | | C16orf62 | 0.037 (9) | | | | ETS1 | | | 0.555 (8) | 0.696 (12) |
| GAB2 | | 0.094 (13) | 0.053 (12) | 0.134 (11) | CBLB* | | 0.057 (8) | | | GALR1 | 0.543 (9) | | 0.699 (15) | 0.848 (13) |
| GOLPH3L | 0.074 (3) | | | | CDH1* | | 0.057 (12) | 0.105 (5) | 0.054 (5) | GSTM2 | 0.383 (5) | 0.755 (5) | | |

*(Continued)*

TABLE 6.1 (CONTINUED)  Driver Gene Prioritization for the Basal-like, Her2, Luminal A and Luminal B Breast Cancer Subtypes

**Copy Number**

| Gene | Basal | Her2 | Luminal A | Luminal B |
|---|---|---|---|---|
| IKBKB | | | 0.053 (6) | 0.116 (15) |
| INTS4 | | 0.094 (12) | 0.057 (8) | 0.152 (5) |
| KCTD14 | 0.025 (13) | 0.094 (9) | 0.057 (5) | 0.143 (7) |
| KDM5A* | 0.049 (6) | | | |
| MCL1 | 0.074 (4) | | | |
| MYC* | 0.185 (1) | 0.245 (2) | 0.086 (3) | 0.179 (3) |
| ORAOV1 | | 0.170 (6) | 0.110 (2) | 0.321 (2) |
| PAK1 | | 0.094 (11) | | |
| PDSS2 | 0.025 (15) | | | |
| QRSL1 | | 0.113 (8) | | |
| RAD52 | 0.037 (8) | | | |
| RPS6KB1 | | 0.170 (4) | | 0.116 (14) |

**Mutation**

| Gene | Basal | Her2 | Luminal A | Luminal B |
|---|---|---|---|---|
| CHGB | | 0.057 (11) | | |
| CTCF | | | 0.048 (8) | |
| DGKG | | | 0.029 (12) | |
| FANCA* | | | | 0.027 (15) |
| GATA3* | | | 0.144 (3) | 0.152 (2) |
| GNPTAB | | 0.057 (9) | | |
| GSPT1 | | | | 0.036 (11) |
| GTSE1 | | | | 0.027 (14) |
| HAUS5 | | 0.057 (6) | | |
| HSPA12A | 0.037 (7) | | | |
| KCNS2 | 0.037 (15) | | | |
| MAP2K4* | | | 0.077 (7) | |

**Methylation**

| Gene | Basal | Her2 | Luminal A | Luminal B |
|---|---|---|---|---|
| JAM3 | 0.432 (4) | 0.811 (2) | 0.555 (6) | 0.670 (9) |
| LAMA1 | | 0.792 (3) | 0.646 (4) | 0.848 (2) |
| ME3 | | 0.604 (12) | | |
| MT1E | 0.407 (13) | | | |
| PRKCQ | | | 0.459 (13) | |
| PXMP4 | 0.235 (15) | | | |
| RIC3 | 0.296 (10) | | | |
| SEMA5A | 0.407 (12) | | | |
| SFRP1 | 0.395 (8) | | 0.589 (7) | 0.741 (7) |
| SLC18A3 | | 0.642 (11) | | |
| SLC22A3 | | 0.811 (1) | 0.780 (1) | 0.839 (1) |
| STK33 | | 0.566 (10) | 0.526 (12) | |

(Continued)

TABLE 6.1 (CONTINUED)  Driver Gene Prioritization for the Basal-like, Her2, Luminal A and Luminal B Breast Cancer Subtypes

| | Copy Number | | | | | Mutation | | | | | Methylation | | | |
|---|---|---|---|---|---|---|---|---|---|---|---|---|---|---|
| | Basal | Her2 | Luminal A | Luminal B | | Basal | Her2 | Luminal A | Luminal B | | Basal | Her2 | Luminal A | Luminal B |
| SUMF2 | 0.037 (10) | | | | MAP3K1 | | | 0.144 (2) | 0.054 (4) | STXBP6 | | | 0.574 (14) | 0.768 (11) |
| THRSP | 0.025 (12) | 0.094 (10) | 0.053 (9) | 0.143 (8) | MLL3* | 0.049 (4) | 0.075 (5) | 0.086 (6) | 0.045 (6) | SV2A | | 0.755 (13) | | |
| TPD52 | 0.037 (11) | | 0.062 (4) | 0.170 (4) | OR6K2 | 0.037 (10) | | | | TRIM58 | 0.654 (3) | | | |
| TUBD1 | | 0.189 (3) | 0.048 (14) | 0.125 (12) | PGLYRP2 | | 0.038 (15) | | | VIM | | | 0.517 (10) | 0.759 (6) |
| VDAC3 | | | 0.053 (7) | | PIK3CA* | 0.037 (13) | 0.264 (2) | 0.306 (1) | 0.143 (3) | VIPR2 | 0.296 (11) | | | 0.625 (15) |
| WNK1 | 0.049 (5) | | | | PIK3R1* | | 0.075 (4) | 0.045 (9) | 0.045 (9) | ZNF671 | 0.765 (2) | | | |
| | | | | | PPEF1 | 0.037 (14) | | 0.019 (14) | | ZCAN12 | 0.370 (14) | 0.887 (8) | 0.641 (11) | 0.821 (8) |
| | | | | | PTEN* | | | 0.043 (10) | 0.045 (8) | | | | | |
| | | | | | RB1* | 0.037 (11) | | | | | | | | |
| | | | | | ROR1 | | 0.057 (13) | | | | | | | |
| | | | | | RPGR | 0.049 (6) | | | | | | | | |
| | | | | | RUNX1* | | | 0.053 (9) | | | | | | |

(Continued)

TABLE 6.1 (CONTINUED)  Driver Gene Prioritization for the Basal-like, Her2, Luminal A and Luminal B Breast Cancer Subtypes

| | Copy Number | | | | | Mutation | | | | Methylation | | | |
|---|---|---|---|---|---|---|---|---|---|---|---|---|---|
| | Basal | Her2 | Luminal A | Luminal B | | Basal | Her2 | Luminal A | Luminal B | Basal | Her2 | Luminal A | Luminal B |
| SEMA5A | | | | | | | | | 0.045 (7) | | | | |
| SETX | | | | | | | | | 0.036 (13) | | | | |
| TBL1XR1 | | | | | | | | 0.024 (15) | | | | | |
| TBX3 | | | | | | | | | 0.045 (10) | | | | |
| TP53* | | | | | | 0.864 (1) | 0.736 (1) | 0.115 (4) | 0.321 (1) | | | | |
| UNC5D | | | | | | 0.062 (3) | | | | | | | |
| USPL1 | | | | | | 0.037 (12) | | | | | | | |
| WASF1 | | | | | | | 0.057 (10) | | | | | | |

*Note:* The top 15 genes with the highest rank, per subtype, for copy number, mutation and methylation data. Values are respectively fraction of patients with altered copy number, mutation frequency and methylation frequency. Values between brackets are the rank of the gene for a given subtype / data type. Genes followed by an asterisk are present in the COSMIC cancer gene census.

different data analysis problems are addressed separately. Sequential data analysis results in information loss and noise propagation as the complementarity between the different data sources cannot be maximally exploited. In addition, such sequential approaches ignore the fact that the different aspects of the tumour analysis problem are intrinsically connected: A subtype is more relevant if patient stratification also highlights the causal drivers and mode of action, and vice versa. This creates opportunities for methods that can integrate heterogeneous system genetics data to comprehensively cope with the tumour analysis problem. As illustrated here, integrative network-based models offer an alternative to sequential analysis. They exploit the potential of their network model by intuitively integrating different data sources, they are ideally suited to identify rare aberrations and they use prior interaction knowledge to gain a mechanistic view on the tumour-related phenotype. A recent analysis in the framework of the ICGC's pan-cancer analysis of whole tumour genomes (PCAWG) illustrated how integrative network-based approaches allow identifying rare non-coding drivers[26] by using the expression information as a proxy for the functional impact of a mutation. Although potentially very powerful, the performance of network-based methods depends highly on the quality and relevance of the network prior. Network priors are intrinsically over connected, because they do not account for tissue/condition dependence. In addition, because most network methods cannot cope with information on the reliability of the prior interactions (edge weights), they rely on highly curated and hence incomplete networks to avoid spurious predictions. Hence fully exploiting the potential of network-based methods requires constructing better network priors.

## Useful Resources

A reference implementation of MUNDIS (with example data) is available for download from http://bioinformatics.intec.ugent.be/mundis.

## References

1. Bailey, M.H. et al. Comprehensive characterization of cancer driver genes and mutations. *Cell* **174**, 1034–1035 (2018).
2. Dimitrakopoulos, C.M. & Beerenwinkel, N. Computational approaches for the identification of cancer genes and pathways. *Wiley Interdisciplinary Reviews. Systems Biology and Medicine* **9**, e1364 (2017).

3. Creixell, P. et al. Pathway and network analysis of cancer genomes. *Nature Methods* **12**, 615–621, 3440 (2015).
4. Cowen, L., Ideker, T., Raphael, B.J. & Sharan, R. Network propagation: A universal amplifier of genetic associations. *Nature Reviews Genetics* **18**, 551–562 (2017).
5. Verbeke, L.P. et al. Pathway relevance ranking for tumor samples through network-based data integration. *PLoS One* **10**, e0133503 (2015).
6. Mizrachi, E. et al. Network-based integration of systems genetics data reveals pathways associated with lignocellulosic biomass accumulation and processing. *Proceedings of the National Academy of Sciences of the United States of America* **114**, 1195–1200 (2017).
7. Dimitrakopoulos, C. et al. Network-based integration of multi-omics data for prioritizing cancer genes. *Bioinformatics* **34**, 2441–2448 (2018).
8. Leiserson, M.D. et al. Pan-cancer network analysis identifies combinations of rare somatic mutations across pathways and protein complexes. *Nature Genetics* **47**, 106–114 (2015).
9. Hofree, M., Shen, J.P., Carter, H., Gross, A. & Ideker, T. Network-based stratification of tumor mutations. *Nature Methods* **10**, 1108–1115 (2013).
10. Le Van, T. et al. Simultaneous discovery of cancer subtypes and subtype features by molecular data integration. *Bioinformatics* **32**, i445–i454 (2016).
11. Mitra, K., Carvunis, A.R., Ramesh, S.K. & Ideker, T. Integrative approaches for finding modular structure in biological networks. *Nature Reviews Genetics* **14**, 719–732 (2013).
12. Horn, H. et al. NetSig: Network-based discovery from cancer genomes. *Nature Methods* **15**, 61–66 (2018).
13. Ciriello, G., Cerami, E., Sander, C. & Schultz, N. Mutual exclusivity analysis identifies oncogenic network modules. *Genome Research* **22**, 398–406 (2012).
14. Babur, O. et al. Systematic identification of cancer driving signalling pathways based on mutual exclusivity of genomic alterations. *Genome Biology* **16**, 45 (2015).
15. Kim, Y.A., Cho, D.Y., Dao, P. & Przytycka, T.M. MEMCover: Integrated analysis of mutual exclusivity and functional network reveals dysregulated pathways across multiple cancer types. *Bioinformatics* **31**, i284–292 (2015).
16. Gao, B., Li, G., Liu, J., Li, Y. & Huang, X. Identification of driver modules in pan-cancer via coordinating coverage and exclusivity. *Oncotarget* **8**, 36115–36126 (2017).
17. Pulido-Tamayo, S., Weytjens, B., De Maeyer, D. & Marchal, K. SSA-ME detection of cancer driver genes using mutual exclusivity by small subnetwork analysis. *Scientific Reports* **6**, 36257 (2016).
18. Swings, T. et al. Network-based identification of adaptive pathways in evolved ethanol-tolerant bacterial populations. *Molecular Biology and Evolution* **34**, 2927–2943 (2017).

19. Ng, S. et al. PARADIGM-SHIFT predicts the function of mutations in multiple cancers using pathway impact analysis. *Bioinformatics* **28**, i640–i646 (2012).
20. Vaske, C.J. et al. Inference of patient-specific pathway activities from multidimensional cancer genomics data using PARADIGM. *Bioinformatics* **26**, i237–245 (2010).
21. Bashashati, A. et al. DriverNet: Uncovering the impact of somatic driver mutations on transcriptional networks in cancer. *Genome Biology* **13**, R124 (2012).
22. Fouss, F., Francoisse, K., Yen, L., Pirotte, A. & Saerens, M. An experimental investigation of kernels on graphs for collaborative recommendation and semisupervised classification. *Neural Networks* **31**, 53–72 (2012).
23. Cancer Genome Atlas Network. Comprehensive molecular portraits of human breast tumors. *Nature* **490**, 61–70 (2012).
24. Veeck, J. et al. Aberrant methylation of the Wnt antagonist SFRP1 in breast cancer is associated with unfavourable prognosis. *Oncogene* **25**, 3479–3488 (2006).
25. Jiang, W. et al. Overexpression of cyclin D1 in rat fibroblasts causes abnormalities in growth control, cell cycle progression and gene expression. *Oncogene* **8**, 3447–3457 (1993).
26. Reyna, M.A. et al. Pathway and network analysis of more than 2,500 whole cancer genomes. *bioRxiv* 385294 (2018).

## 6.2 PATIENT STRATIFICATION FROM SOMATIC MUTATIONS
*Jean-Philippe Vert*

### 6.2.1 Summary
The transition from a normal cell to a cancer cell is driven by genetic alterations, such as mutations, that induce uncontrolled cell proliferation. With
the advent of next-generation sequencing technologies (NGS) in the last
decade, thousands of tumours have been sequenced and their mutation profiles determined. However, the statistical analysis of these mutation profiles
is challenging. First, two patients rarely share the same set of mutations, and
can even have none in common. Second, it is difficult to distinguish the few
disease-causing mutations from the dozens, often hundreds of mutations
observed in a tumour. To alleviate these challenges, it has been proposed to
use gene–gene interaction networks as prior knowledge, with the idea that if a
gene is mutated and becomes non-functional, then its interacting neighbours
might not be able to fulfil their function as well. Here we describe NetNorM,
a method proposed by Le Morvan et al.,[1] that transforms mutation data using
gene networks so as to make mutation profiles more amenable to statistical
learning. NetNorM was shown to significantly improve the prognostic power
of somatic mutation data compared to previous approaches, and to allow
defining meaningful groups of patients based on their mutation profiles.

### 6.2.2 Introduction
Tumourigenesis involves somatic mutations which appear and accumulate
during cancer growth. These mutations impair the normal behaviour of various cancer genes, and give cancer cells an advantage to proliferate over normal cells.[2–4] Systematically assessing and monitoring somatic mutations in
cancer therefore offers the opportunity to help rationalize patient treatment
in a clinical setting. This involves finely characterising the genomic abnormalities of each patient to discover which may be treatable by a targeted therapeutic agent, as well as improving prognosis using molecular information.[5–7]

We now have access to collections of thousands of sequenced tumours.[8,9]
By comparing DNA in tumour and normal cells, we can systematically
identify somatic mutations in each tumour, i.e. mutations present in
the tumour but not in the corresponding normal tissue. Mapping back
the somatic mutations to the genes they affect allows the creation of the
*somatic mutation profile* of each tumour, i.e. the list of genes which are
affected by mutations in the tumour. Stratifying patients based on the

somatic mutation profile of their tumour may then lead to new, genomic-based groups relevant to predict the response to treatment or survival.[10–13]

The comparison and analysis of somatic mutation profiles is however challenging for multiple reasons. First, most somatic mutations detected by systematic sequencing are so-called *passenger* mutations, irrelevant for clinical applications.[4,14] Second, sequencing efforts have shown that while a few genes are frequently mutated, the vast majority of genes are mutated in only a handful of patients.[15,16] As a result, the mutation profiles of two tumours often only share a few if any genes in common. Third, even if originating from the same tissue, tumours may exhibit widely varying mutation rates. The overall mutational burden of a tumour constitutes a strong and informative signal[17–19] but can complicate the retrieval of more subtle signals. Combined with the inherent high dimensionality of somatic mutation datasets, this makes any statistical analysis of cohorts of whole-exome somatic mutation profiles extremely challenging.

Here we describe NetNorM,[1] a computational method to address these issues. NetNorM uses a gene network as prior knowledge to normalize somatic mutation profiles and make them more amenable to comparison and statistical analysis, an idea already proposed and investigated by Hofree et al.[20] We show that NetNorM normalization improves survival prediction and patient stratification in some cases, and refer the interested reader to the original publication[1] for more details.

### 6.2.3 Approach and Application Example

Let us first present NetNorM, borrowing the description in the original publication.[1] NetNorM is a computational method that takes as input an undirected gene network and raw exome somatic mutation profiles and outputs a new representation of mutation profiles which allows better survival prediction and patient stratification from mutations. Here and in what follows, the 'raw' mutation profiles refer to the binary patients times genes matrix where 1s indicate non-silent somatic point mutations or indels in a patient-gene pair and 0s indicate the absence of such mutations. The new representation of mutation profiles computed with NetNorM also takes the form of a binary patients times genes mutation matrix, yet with new properties. While different tumours usually harbour different number of mutations, with NetNorM all patient mutation profiles are normalized to the same number $k$ of genes marked as mutated. The final number of mutations $k$ is the only parameter of NetNorM which can be adjusted by various heuristics, such as the median number of mutations in the original profiles, or optimized by cross-validation for a given task such as survival prediction. In order to represent each tumour by $k$ mutations, NetNorM adds 'missing' mutations to samples with less than $k$ mutations, and removes 'non-essential' mutations from samples with more than $k$ mutations. The 'missing' mutations added to a sample with few mutations are the non-mutated genes with the largest number of mutated neighbours in the gene network, while the 'non-essential' mutations removed from samples with many mutations are the ones with the smallest degree in the gene network. These choices rely on the simple ideas that, on the one hand, genes with a lot of interacting neighbours mutated might be unable to fulfil their functions and, on the other hand, mutations in genes with a small number of interacting neighbours might have a minor impact compared to mutations in more connected genes. We refer the interested reader to Le Morvan et al.[1] for more details about NetNorM (Figure 6.2).

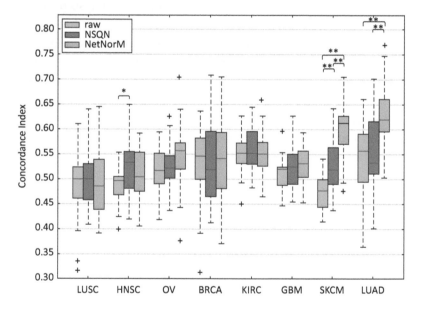

FIGURE 6.2 Comparison of the survival predictive power of the raw mutation data, NSQN and NetNorM (with Pathway Commons as gene network) for eight cancer types. For each cancer type, samples were split 20 times in training and test sets (four times five-fold cross-validation). Each time a sparse survival SVM was trained on the training set and the test set was used for performance evaluation. The presence of asterisks indicate when the test CI is significantly different between two conditions (Wilcoxon signed-rank test, $p < 5 \times 10{-}2$ [*] or $p < 1 \times 10{-}2$ [**]). (Figure and legend from Le Morvan M., Zinovyev, A., Vert, J-P. NetNorM: Capturing cancer-relevant information in somatic exome mutation data with gene networks for cancer stratification and prognosis. *PLoS Comput Biol.* **13**(6), e1005573. [2017].)

To assess the benefits of NetNorM, we evaluated its performance and compared it to alternative approaches in several tasks of patient stratification and survival prediction, using data from The Cancer Genome Atlas (TCGA). Here we just focus on survival prediction, where we want to predict how long a patient is likely to survive after diagnosis, given the mutation profile of the tumour. We collected a total of 3,278 full-exome mutation profiles of eight cancer types from the TCGA portal, together with survival information and clinical data: lung adenocarcinoma (LUAD, N = 430), skin cutaneous melanoma (SKCM, N = 307), glioblastoma multiform (GBM, N = 265), breast invasive carcinoma (BRCA, N = 945), kidney renal clear cell carcinoma (KIRC, N = 411), head and neck squamous cell carcinoma (HNSC, N = 388), lung squamous cell carcinoma (N = 169), ovarian serous cystadenocarcinoma (OV, 363). In parallel, since NetNorM is based on a gene network as prior knowledge, we retrieved a gene network connecting 16,674 human genes through more than two million edges from the Pathway Commons database, which integrates a number of pathway and molecular interaction databases. To evaluate the performance of a method for survival prediction, we performed a cross-validation experiment where for each cancer type (e.g. breast carcinoma), 1) we randomly split the set of TCGA samples into an 80% training set and a 20% test set; 2) we train a model for survival prediction on the training set, using both the somatic mutation profiles and the survival information for the corresponding samples; 3) we use the resulting model to predict survival on the test samples, given only their somatic mutation profiles; 4) we compare the predicted versus known survival data on the test set, quantifying the similarity with the concordance index (CI), a variant of correlation adapted to censored survival information data. A random predictor should have CI = 0.5, while a perfect predictor should have CI = 1. We repeated the whole procedure 20 times with different train/test splits, and assessed the performance of a method through the distribution of CI on the test set. Since our goal was to assess the influence of the original representation of the somatic mutation profile (i.e. using directly the raw binary profile, or the one normalized by NetNorM or by the network smoothing method of Hofree et al.[20]), we also established a procedure to train a survival model given any representation of the somatic mutation profile: We used a state-of-the-art sparse survival SVM with parameters automatically tuned on the training set by internal cross-validation. We refer the interested reader to the original publication and the online code for more details.

Figure 6.1 summarizes the performance of using the raw mutation profile, or its normalized version by NetNorM or NSQN (for network smoothing and quantile normalization, which refers here to the method of Hofree et al.[20]), for survival prediction on eight cancer types. We see that for two cancers (LUSC, HNSC), none of the methods manages to outperform a random prediction, questioning the relevance of the mutation information in this context. For OV, BRCA, KIRC and GBM, all three methods are significantly better than random, although the estimated CI remains below 0.56, and we again observe no significant difference between the raw data and the data transformed by NSQN or NetNorM. Finally, the last two cases, SKCM and LUAD, are the only ones for which we reach a median CI above 0.6. In both cases, processing the mutation data with NetNorM significantly improves performances compared to using the raw data or profiles processed with NSQN. More precisely, for LUAD the median CI increases from 0.56 for the raw data and 0.53 for NSQN to 0.62 for NetNorM. In the case of SKCM, the median CI increases from 0.48 for the raw data to 0.52 for NSQN, and to 0.61 for NetNorM. For SKCM, both NetNorM and NSQN are significantly better than the raw data ($p < 0.01$).

These results suggest that the initial processing of the raw somatic mutation profiles can impact the performance of statistical models fitted on them, and that in the cases where it does, NetNorM provides an advantage compared to both the raw data, and the raw data normalized by the network-based smoothing technique NSQN. In the NetNorM publication[1] we further investigate various properties of NetNorM. By randomly shuffling the nodes of the biological network, we demonstrate that the performance decreases, demonstrating that NetNorM's performance uses biological information encoded in the network beyond its topology. By analyzing the genes selected by the sparse survival SVM model, we show that NetNorM captures both specific mutations known to be important in some cancers (such as the negative prognosis brought by a mutation in TP53 in LUAD), and pathway-level information capturing the frequency of mutations in some specific areas of the network.

## 6.2.4 Discussion and Perspectives

Exploiting the wealth of cancer genomic data collected by large-scale sequencing efforts is a pressing need for clinical applications. Somatic mutations are particularly important since they may reveal the unique history of each tumour at the molecular level, and shed light on the biological processes and potential drug targets dysregulated in each patient. Standard statistical techniques for unsupervised classification or supervised predictive modelling perform poorly when each patient is represented by a raw binary vector indicating which genes have a somatic mutation. This is both because the relevant driver mutations are hidden in the middle of many irrelevant passenger mutations, and because there is usually very little overlap between the somatic mutation profiles of two individuals. NetNorM aims to increase the relevance of mutation data for various tasks such as prognostic modelling and patient stratification by leveraging gene networks as prior knowledge.

Besides the specificities of the NetNorM algorithm, we see this study as an illustration that modelling and data representation still play an important role in the 'big data' era. As new technologies allow measuring of an increasing amount of information in a medical context (including genomic information, but also images or medical records), the description of each patient gets richer and richer, but the size of cohorts available to fit statistical and machine learning models can not keep up with the pace of growth. Hence we need more than ever to develop methods and approaches to exploit rich information from a limited number of samples. NetNorM is one among many approaches to apply this principle to the study of cancer mutation profiles. The increased performance it brings shows that this is not only a theoretical consideration, but that it can translate to actual improvement in survival prediction.

## Useful Resources

https://github.com/marineLM/NetNorM: code and information to use NetNorM and reproduce all experiments in the publication[1].

## References

1. Le Morvan, M., Zinovyev, A., Vert, J-P. NetNorM: Capturing cancer-relevant information in somatic exome mutation data with gene networks for cancer stratification and prognosis. *PLoS Comput Biol.* **13**(6), e1005573. (2017).
2. Stratton, M.R., Campbell, P.J., Futreal, P.A. The cancer genome. *Nature* **458**, 719–724. 2009.
3. Hanahan, D., Weinberg, R.A. Hallmarks of cancer: The next generation. *Cell.* **144**(5), 646–674. (2011).
4. Vogelstein, B., Papadopoulos, N., Velculescu, V.E., Zhou, S., Diaz, L.A. Jr, Kinzler, K.W. Cancer genome landscapes. *Science* **339**(6127), 1546–1558. (2013).

5. Chin, L., Gray, J.W. Translating insights from the cancer genome into clinical practice. *Nature* **452**(7187), 553–563. (2008).

6. Olivier, M., Taniere, P. Somatic mutations in cancer prognosis and prediction: Lessons from TP53 and EGFR genes. *Curr Opin Oncol.* **23**(1), 88–92. (2011).

7. Mardis, E.R. Genome sequencing and cancer. *Curr Opin Genet Dev.* **22**(3), 245–250. (2012).

8. The Cancer Genome Atlas Research Network, Weinstein, J.N., Collisson, E.A., Mills, G.B., Shaw, K.R.M, Ozenberger, B.A. et al. The Cancer Genome Atlas Pan-Cancer analysis project. *Nat Genet.* **45**(10), 1113–1120. (2013).

9. Hudson, T.J., Anderson, W., Aretz, A., Barker, A.D. International network of cancer genome projects. *Nature* **464**(7291), 993–998. (2010).

10. The Cancer Genome Atlas. Comprehensive genomic characterization defines human glioblastoma genes and core pathways. *Nature* **455**(7216), 1061–1068. (2008)

11. The Cancer Genome Atlas Research Network. Integrated genomic analyses of ovarian carcinoma. *Nature* **474**(7353), 609–615. (2011).

12. The Cancer Genome Atlas Network. Comprehensive molecular portraits of human breast tumors. *Nature* **490**(7418), 61–70. (2012).

13. Kandoth, C., McLellan, M.D., Vandin, F., Ye, K., Niu, B., Lu, C. et al. Mutational landscape and significance across 12 major cancer types. *Nature* **503**(7471), 333–339. (2013).

14. Greenman, C., Stephens, P., Smith, R., Dalgliesh, G.L., Hunter, C., Bignell, G. et al. Patterns of somatic mutation in human cancer genomes. *Nature* **446**(7132), 153–158. (2007).

15. Wood, L.D., Parsons, D.W., Jones, S., Lin, J., Sjöblom, T., Leary, R.J. et al. The genomic landscapes of human breast and colorectal cancers. *Science* **318**(5853), 1108–1113. (2007).

16. Lawrence, M.S., Stojanov, P., Mermel, C.H., Robinson, J.T., Garraway, L.A., Golub, T.R. et al. Discovery and saturation analysis of cancer genes across 21 tumor types. *Nature* **505**(7484), 495–501. (2014).

17. Lawrence, M.S., Stojanov, P., Polak, P., Kryukov, G.V., Cibulskis, K., Sivachenko, A. et al. Mutational heterogeneity in cancer and the search for new cancer-associated genes. *Nature* **499**(7457), 214–218. (2013).

18. Birkbak, N.J., Kochupurakkal, B., Izarzugaza, J.M.G., Eklund, A.C., Li, Y., Liu, J. et al. Tumor mutation burden forecasts outcome in ovarian cancer with BRCA1 or BRCA2 mutations. *PLoS One* **8**(11), e80023. (2013)

19. Rizvi, N.A., Hellmann, M.D., Snyder, A., Kvistborg, P., Makarov, V., Havel, J.J. et al. Mutational landscape determines sensitivity to PD-1 blockade in non-small cell lung cancer. *Science* **348**(6230), 124–128. (2015).

20. Hofree, M., Shen, J.P., Carter, H., Gross, A., Ideker, T. Network-based stratification of tumor mutations. *Nat Methods* **10**(11), 1108–1115. (2013).

## 6.3 EVALUATING GROWTH AND RISK OF RELAPSE OF INTRACRANIAL TUMOURS

*Olivier Saut, Thierry Colin, Annabelle Collin, Thibaut Kritter, Vivien Pianet, Clair Poignard and Benjamin Taton*

### 6.3.1 Summary

As cancer evolution is challenging to evaluate, there is dire need of novel approaches offering clinicians a better insight on the disease. For instance, having an estimation of the growth of slowly evolving tumours that have to be monitored or of the risk of relapse after treatment may be invaluable for clinicians. In this chapter, two approaches (statistical learning and mechanistic modelling) are presented that aim at addressing these clinical questions. As we wish to use data available in the clinical routine for solid tumours, medical images will be a major source of insight on the disease.

### 6.3.2 Introduction

In order to evaluate the evolution of a tumour, different approaches may be used depending on the type of data available. When longitudinal data are available, they can be used to build a mechanistic model of the evolution of this disease. This model could then give a quantitative estimate or a prediction of this evolution. When only one single time point is available, statistical approaches are more suited. One tries to correlate different observed features of the disease with clinical outcome by using statistical learning techniques. *In this article, these two different kinds of approaches are presented through two examples from a clinical context.*

In the first example, we show how a mechanistic model describing the growth of meningioma (a type of intracranial tumour) may be useful to have an estimation of the future size, volume, location of the targeted lesion. This information could help clinicians adjust a patient's follow-up or plan a surgery if a specific area of the brain is at risk of being deformed by the increase of volume of the tumour. The derivation of mechanistic models is based on the biological knowledge and the available observations. Mathematical models have become popular tools to analyze longitudinal data in oncology where we want to obtain patient-specific prognoses for which purely statistical models are not adequate. The earliest models described the evolution of tumour diameters on mice for preclinical studies. These models – e.g. those based on Gompertz's law – have few parameters and were shown adequate to fit experimental data and in some cases, even have a certain predictive value. However, when coupled with imaging data, these models neglect most spatial information even if they describe tumour area or volume and not only the diameter of the lesion. To fully exploit medical images, it is therefore key to develop mathematical models describing the spatial evolution of the tumour at the scale of these images (which makes them essentially phenomenological). These allow more information to be kept from the data and one can expect to have a better insight on the tumour.

In our second example, we show how radiomics approaches[1] may be helpful in evaluating the risk of relapse of low-grade gliomas. This example illustrates how to combine information at different scales to obtain this evaluation. Whereas the former approach studied each patient individually (by personalizing a model), this approach uses the whole cohort to determine a statistical correlation between various clinical, omics and imaging features and the clinical outcome of the patient. A significant number of patients is required in order to identify a significant correlation.

### 6.3.3 Approach and Application Example

Meningioma are rather benign intracranial tumours. One major clinical challenge is to evaluate their growth to help surgeons deciding if they should be resected (as the intervention may be risky depending on the location of the lesion) or carefully monitored. More precisely, the question we are trying to address is the following one: Given two MRIs of a patient where the meningioma has been delineated, are we able to predict the location, shape and volume of the lesion at a given later time?

Different approaches are possible to model spatial tumour growth. Basically, two large classes of models may be distinguished. In the so-called *discrete model class*, the evolution of each cancer or healthy cell is described individually (hence at the microscopic scale). Agent-based models (like in[2]) describe the evolution of individual cells. This class of models is well adapted to describe the smaller scales. However, these models are computationally very expensive and rendering mechanical effects might be difficult (for instance, one has to consider the interaction of each cell with its neighbours). On the other hand, in *continuous or macroscopic models*, one describes densities of cells, i.e. averages over a large number of cells or typically in a voxel of a medical image. Voxel-based models are often based on a set of partial differential equations (PDE) and describe cellular densities in each voxel as well as – in some cases – nutrients concentration or blood flow. This is the class of models we use for describing meningioma.

In this study in collaboration with CHU Bordeaux, meningioma are not treated and are growing naturally. In close collaboration with clinicians, key features of the disease (the tumour grows slowly from the arachnoid, is rather homogeneous) were translated into a mechanistic model based on a set of partial differential equations[3]. This simple model describes the evolution of the density of cancer cells and of a growth factor that is decreasing over time. By dividing, cancer cells push their neighbours and create a global passive movement that is described by a velocity assumed to follow Darcy's law. This model has two parameters that are patient specific.

A key factor to the use of mathematical models for clinical applications is the recovery of their parameters. Indeed, several parameters involved in these models do not have any physical meaning nor can be measured experimentally. A way has to be found to overcome this difficulty. Once they are recovered, one can run the mathematical model and compute the future evolution of the tumour. If the mathematical model is biologically accurate, this computed evolution will not be much different from the real one and can be seen as a prognosis.

Here, the main source of information on the disease is obtained from images. From these images, we use the delineated tumour and compared its evolution with the simulation run with our model. In our case, the model can be integrated and the system of PDE transformed into an equation describing the evolution of the tumour volume. This equation has the same parameters as the full model. A Bayesian technique is then used to recover reasonable values of the parameters. The personalized model is then validated on a retrospective study from the Bordeaux Hospital with very satisfactory accuracy[3]. Yet this technique does not work for tumours with complex shape evolutions (like brain metastases or gliomas). For this matter, more advanced data assimilation techniques have to be used like sequential approaches and state observers (that may also account for uncertainties in the observations). We have recently extended the approach of[4] with very promising results.

On the other hand, when no longitudinal information is available, other techniques have to be used. For instance, in collaboration with Humanitas Research Hospital, we tried to evaluate the risk of relapse of patients with low-grade gliomas (a type of brain tumour). These patients had their tumour resected and the challenge for the clinicians is to determine whether they will relapse before or after 30 months (in order to adapt their follow-up). For a cohort of more than 100 patients, we had clinical information (sex, age, etc.), histo-molecular data (IDH1 status, codeletion) as well as information extracted from images (PET-MET scans and MRIs). From the images we compute various features (tumour volume, enhancement volume, SUVMax, SUVMean, etc.) as well as a novel PET heterogeneity marker described in[5]. For each patient we know her/his clinical outcome, so we can determine if a relapse occurred before or after 30 months.

On this data, we trained a statistical learning algorithm to correlate the features to clinical outcome. This is achieved with a ten-fold cross-validation. The number of features is reduced through standard feature selection and dimensionality reduction algorithms. Several classical algorithms are then tested. With the best classifier the mean accuracy is around 82% and the AUC is at 0.82 which is very satisfactory. In particular, as shown in Figure 6.3, it gives a better stratification of patients than a classification by histological grade or by the IDH1 status (which is not

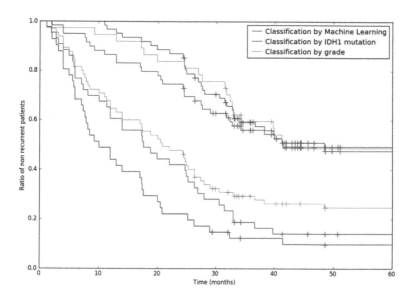

FIGURE 6.3   Kaplain–Meier curves on the cohort from Humanitas Hospital (in collaboration with L. Bello and M. Rossi). Three different stratifications are evaluated: by grade (where risk of relapse is directly related to grade), by IDH1 status (where we split patient regarding the IDH1 mutation status) and by our machine learning algorithm (which uses grade, IDH1 status as well as other clinical, omic or imaging features). Our approach yields a much better stratification of patients.

entirely striking as our algorithm uses these features – and others – to discriminate patients).

## 6.3.4 Discussion and Perspectives

We have presented two different approaches to obtain a better evaluation of the evolution of some intracranial tumours. The first one is purely patient-specific and relies on a mechanistic model that is personalized for each patient. The second one relies on a patient cohort and statistics to evaluate a risk of relapse. Both approaches are well adapted to the challenges they are trying to overcome. They show that mathematics could offer meaningful insights for clinicians in their routine.

Yet there are several shortcomings that may prevent the use of these approaches in other pathologies or contexts. There is obviously no generic model for cancer progression or response to therapy. A novel model has to be developed for each type of cancer targeted *and* the available data. This requires a strong collaboration between mathematicians, computer scientists, biologists and clinicians. Given the complexity of the cancer mechanisms, modelling is always a trade-off between accuracy (precise description of the biology) and applicability (the effective use of the model for our applications).

The personalization process takes time as it requires exploring the parameter space[6]. Yet this calibration could be shortened with reduced-order models derived from the original models. These models – that are much faster to compute that the ones from which they are build – reduce the time taken by data assimilation algorithms that require many evaluations of the model for different sets of parameters. These reduced-order models may for instance be built analytically (as in the presented example), through proper-orthogonal decomposition[7] or with machine learning techniques.

A promising perspective is the development of hybrid approaches combining cohort information and individual evolutions. For simpler mechanistic models (based on ordinary differential equations), mixed-effect models can be used but their extension to spatial models based on partial differential equations is far from being straightforward[8]. Another approach, that we are currently investigating, is to include model parameters as patient-specific features in a machine-learning approach.

Finally, as the ultimate goal is to develop decision-helping tools for clinicians, great care should be taken to ensure the robustness of these tools. The answer they offer should not be sensitive to noise and uncertainty (even in delineations[9]), model prediction[10], data variability (between medical centres and medical devices) for instance. This has to be investigated seriously for any useful clinical application.

## References

1. Parmar, C., Grossmann, P., Bussink, J., Lambin, P. & Aerts, H.J. Machine learning methods for quantitative radiomic biomarkers. *Scientific Reports* **5**, 13087. (2015).
2. Mansury Y., Kimura M., Lobo J., Deisboeck T.S. Emerging patterns in tumor systems: Simulating the dynamics of multicellular clusters with an agent-based spatial agglomeration model. *Journal of Theoretical Biology* **219**(3), 343–370. (2002).
3. Pianet, V, Colin, T., Loiseau, H., Joie, J., Lafourcade, J., Kantor, G., Bigourdan, A., Taton, B. & Saut, O. P12.23 tumor growth model applied for meningiomas: First clinical validation. *Neuro-Oncology* **19**, 99–100. (2017).
4. Collin, A., Chapelle, D. & Moireau, P. A Luenberger observer for reaction–diffusion models with front position data. *Journal of Computational Physics* **300**, 20. (2015).
5. Kritter, T., Rossi, M., Colin, T., Cornelis, F.H., Egesta, L., Bello, L., Chiti, A., Poignard, C. & Saut, O. Heterogeneity index: A methionine PET based prognostic factor in low grade gliomas, submitted.
6. Lê, M., Delingette, H., Kalpathy-Cramer, J., Gerstner, E., Batchelor, T., Unkelbach, J. & Ayache, N. MRI based bayesian personalization of a tumor growth model. *IEEE Transactions on Medical Imaging* **35**(10), 2329–2339. (2016).
7. Colin, T., Cornelis, F., Jouganous, J., Palussière, J. & Saut, O. Patient specific simulation of tumor growth, response to the treatment and relapse of a lung metastasis: A clinical case. *Journal of Computational Surgery* **2**, 1. (2015).
8. Grenier, E., Louvet, V. & Vigneaux, P. Parameter estimation in non-linear mixed effects models with SAEM algorithm: Extension from ODE to PDE. *ESAIM: Mathematical Modelling and Numerical Analysis* **48**(5), 1303–1329. (2014).
9. Cornelis, F.H., Martin, M., Saut, O., Buy, X., Kind, M., Palussiere, J. & Colin, T. Precision of manual two-dimensional segmentations of lung and liver metastases and its impact on tumor response assessment using RECIST 1.1. *European Radiology Experimental* **1**, 16. (2017).
10. Hawkins-Daarud, A., Johnston, S. K. & Swanson, K. R. Quantifying uncertainty and robustness in a biomathematical model based patient-specific response metric for glioblastoma. *bioRxiv*. (2018).

## 6.4 MACHINE LEARNING FOR SYSTEMS MICROSCOPY

*Thomas Walter*

### 6.4.1 Summary

Today, we have the technologies and computational tools to analyze entire genomes, transcriptomes and proteomes and thereby obtain unprecedented insights into the molecular basis of living systems in general and disease in particular. Yet, in order to truly understand the complex genotype–phenotype relationships, it is also necessary to study the phenotype of the system as comprehensively as possible.

Large-scale imaging approaches, as provided by High Content Screening, are well suited to study cellular and tissular phenotypes and to thereby fill this gap. In order to analyze the massive and often complex image datasets generated by these approaches, machine learning provides a powerful set of techniques. In this chapter, we will show the evolution of the field and discuss different methodological options.

### 6.4.2 Introduction

In the last decades, technological developments, such as next-generation sequencing, have triggered revolutionary changes in the life sciences. In particular, the many new technologies have reinforced the quantitative aspects of biology and encouraged ambitious, large-scale projects involving a large number of research institutions worldwide, aiming at understanding life at the systems level. Importantly, the availability of large amounts of systematically acquired high-quality data is at the very heart of this shift in paradigm, thus transforming biology – to a large extent – into a *data science*. This shift in focus has resulted in an ever-increasing importance of the development of sophisticated and robust computational methods and tools in *data mining, statistics* and *machine learning* to take best advantage of these massive amounts of heterogeneous and complex data.

Long before these technical breakthroughs, imaging has been and still is one of the most important and most widely used experimental techniques in biology. Importantly, imaging approaches are complementary to most molecular approaches in several ways. In particular, they allow one:

- To explore the *spatial dimension* of living systems, such as the spatial arrangements of proteins inside cells, or the organization of cells in a developing organism.

- To explore the *temporal dimension*. Microscopy is certainly among the most efficient techniques when it comes to studying the behaviour of biological systems over time.

- To study living systems at *different scales of organization*: The molecular, cellular, tissular and organism scale are accessible to imaging approaches. In many cases, we can even have access to two or more of these scales at the same time.

- To investigate the morphological properties of biological entities (cells, organisms). In particular, this allows the investigation of the effect of some treatment on the general *phenotype* of a cell or organism.

In the last decade, technical advances in the field of imaging techniques, microscopy automation and the availability of large-scale libraries of reagents allow for the generation of extremely large image datasets,[1] that match both size and complexity of typical omics datasets. As these approaches allow for comprehensive phenotypic screening, with each single experiment containing rich information on the biological system under study, they are referred to as *high content screening*,[1,2] which is the major workhorse in the field of *systems microscopy*. This technical evolution has given rise to the new discipline bioimage informatics,[3-7] with a particular focus on machine learning and computer vision methods to unravel the phenotypic complexity of living systems.

## 6.4.3 Approach and Application Example

Imaging approaches are in general well suited to the study of phenotypes, as they give access to a large number of fundamental properties of cells, tissues and organisms, while maintaining their spatial organization and morphological integrity. The difficulty in the computational analysis of such data is thus to capture all of the interesting properties, to describe them quantitatively and to infer biological knowledge from these descriptors.

The first step of these methods usually consists of segmenting cells or cellular compartments from the images. This step can be rather trivial or exceedingly complicated, depending on the objects that are studied, the performance we need to achieve and the imaging modalities and markers that are used. In the context of fluorescence microscopy, we usually start by segmenting the cell nuclei in a first step, as they are informative on the cellular state and many important phenotypes. Importantly, they are usually easier to segment than most other compartments of the cell. Segmentation is its own field of research and there exists a plethora of methods and tools to achieve segmentation of nuclei, cells and cellular compartments for both cultured cells[8,9] or more challenging imaging modalities, such as stained tissue sections or electron microscopy data,[10,11] which we will not discuss further in this overview chapter.

In the conceptually easiest case, there is one feature that fully describes the biological process under study (Figure 6.4A). Obvious examples of such features include cell size, overall fluorescence intensity or ratio measurements at specific locations in the cell (e.g. in the DAPI channel as a proxy for the cell cycle phase).[12,13] A cellular population is thus described by one univariate distribution of feature values. While limited in many aspects, it is the simplicity in the phenotype description by a simple feature

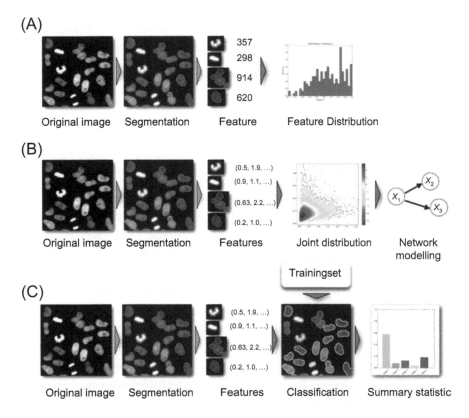

FIGURE 6.4 Different methodological choices for the analysis of large-scale screening data: (A) Univariate feature distribution, (B) Multivariate feature distribution, (C) Object classification approach.

that makes the subsequent analysis steps more straightforward and that explains the popularity of the strategy.

In most cases, there is more than one relevant feature that can be measured for a cell. Indeed, even if it is possible to quantitatively describe the biological process under study with just one feature (as described earlier), it is usually interesting to relate this feature to other features measured for the same cells. Hence, each cell is described by a vector of feature values, covering different aspects of the phenotype. The population of cells is thus described by the joint distribution of feature values, which can be further analyzed by graphical modelling[14,15] or causal inference techniques.[16] This is illustrated in Figure 6.4B.

Both the univariate and the multivariate feature approach rely on the existence of interpretable features, i.e. image features that correspond to interpretable biological properties of the cells. Many visually noticeable differences, however, do not relate to a single interpretable feature, but might require several rather abstract texture or shape features to be properly accounted for. In such a case, the strategy is to compute a large number of potentially non-interpretable features describing the segmented objects. Then, there are two scenarios:

- *Supervised Learning:* In this case, we define a set of phenotypic classes and provide example objects for each of these classes. The union of all of these annotated samples builds the training set and represents our prior knowledge on the phenotypes we wish to identify. Then we train a classifier on the training data that allows us to predict the phenotypic class for unseen data from the feature representation. The advantage of supervised learning is that we can guarantee that the output is biologically meaningful, as we predefine the classes according to our biological knowledge on the biological system. The drawback is that we might not know all the phenotypic classes in a large-scale dataset in advance, and for this reason novelty detection (at the single cell level) is not possible in this setting. Methods used in this context include linear discriminant analysis (LDA), support vector machines (SVM),[17,18] gentle boosting[19] or random forests (RF).[20] More recently, convolutional neural networks (CNN) have also been used for single cell classification,[21] thereby omitting the feature design step, but usually requiring more manual annotation.

- *Unsupervised Learning:* In this case, there are no annotations, and the classes are inferred from the data. The advantage is that this analysis is unbiased and allows for the detection of new phenotypes. In addition to being computationally challenging for large amounts of data, the drawback of unsupervised learning is that the biological sense is not guaranteed: Two cells might be visually different but biologically very close and vice versa. In contrast to supervised learning, there is no way of imposing 'biological meaningfulness'. Methods for unsupervised learning in this context include methods for dimensionality reduction, like PCA[22] or t-SNE,[23] and clustering methods, such as k-means.[24]

- Given the individual classification results for each cell, the cellular population is then described by a vector of classification results or a summary statistic thereof, typically the percentages of cells in all classes. This approach is illustrated in Figure 6.4C.

*Examples:* High Content Screening has been a trigger for the use of machine learning to microscopy data. Indeed, the first large-scale applications of cell classification were in the field of protein localization, where the objective was to investigate the subcellular localization of proteins. As individual proteins can usually not be resolved, the patterns were described by general texture features.[25-27] Interestingly, similar or identical features could also be used to classify morphological phenotypes, complemented by shape features which are irrelevant for the classification of localization patterns.[15,17] Morphological phenotyping by machine learning has also been applied to large-scale live cell imaging data[18] to study the morphological changes of cells and cellular compartments over time. Also, phenotyping is not limited to morphologies and localization patterns: Trajectories can also be described by dedicated movement features, and with unsupervised learning techniques, we can identify the different movement types in a set of live cell imaging experiments.[28] One of the most recent applications of machine learning to large-scale screening data deals with the spatial aspects of gene expression, which can be studied by single molecule in situ hybridization (smFISH). With this technique, we can visualize individual RNA molecules in cells and tissues. In contrast to protein localization assays, we can resolve individual molecules and thereby represent each cell by a point cloud, that can be described with features from spatial statistics. As for this type of data it is more difficult to obtain robust manual annotations, it is also possible to train classifiers from simulated microscopy data,[23] opening up interesting perspectives to link physical and statistical models.

### 6.4.4 Discussion and Perspectives

While machine learning for microscopy data was mainly used in the context of large-scale screening in the past, it has gained a lot of popularity over the last number of years. There are domains of traditional image analysis that are literally about to disappear and be replaced by learning approaches, even though this was not imaginable ten years ago.

Today, the most widely used machine learning technique applied to images is deep learning, or more precisely, convolutional neural networks (CNN). This technique has the advantage of learning representations of the data in addition to the class labels, i.e. we do not need to define features a priori. This huge benefit is hampered by the number of annotations typically required to train these models. Much of the research today in this field deals with the question of how to generate large annotated datasets. Different strategies include crowd-sourcing and gamification,[29] where annotations are disguised in video games, image simulation where arbitrary amounts of data can be generated to build very large datasets[23] with the additional difficulty that the simulated data does not necessarily follow the same distribution as real data, or the generation of experimental ground truth,[30] where no manual annotation is required as the annotation is provided by a smart experimental setup. Importantly, deep learning can also be used for image generation, which is a particularly popular and promising strategy. Indeed, today we can use deep learning to predict super-resolved images or additional fluorescent markers from low-resolution and non-invasive imaging techniques, respectively. Deep learning is therefore not only improving classification accuracies, but really opens new avenues in bioimage analysis.

### References

1. Pepperkok, R. & Ellenberg, J. High-throughput fluorescence microscopy for systems biology. *Nat. Rev. Mol. Cell Biol.* **7**, 690–696 (2006).
2. Carpenter, A.E. & Sabatini, D.M. Systematic genome-wide screens of gene function. *Nat. Rev. Genet.* **5**, 11–22 (2004).
3. Danuser, G. Computer vision in cell biology. *Cell* **147**, 973–978 (2011).
4. Myers, G. Why bioimage informatics matters. *Nat. Methods* **9**, 659–660 (2012).

5. Peng, H., Bateman, A., Valencia, A. & Wren, J.D. Bioimage informatics: A new category in Bioinformatics. *Bioinformatics* **28**, 1057 (2012).
6. Cardona, A. & Tomancak, P. Current challenges in open-source bioimage informatics. *Nat. Methods* **9**, 661–665 (2012).
7. Coelho, L.P. et al. Principles of bioimage informatics: Focus on machine learning of cell patterns. *Lect. Notes Comput. Sci.* **6004**, 8–18 (2010).
8. Meijering, E. Cell segmentation: 50 years down the road. *IEEE Signal Process. Mag.* **29**, 140–145 (2012).
9. Jones, T.R., Carpenter, A. & Golland, P. Voronoi-based segmentation of cells on image manifolds. *Comput. Vis. Biomed. Image Appl. LNCS* **3765**, 535–543 (2005).
10. Ronneberger, O., Fischer, P. & Brox, T. U-Net: Convolutional networks for biomedical image segmentation. In *Medical Image Computing and Computer-Assisted Intervention – MICCAI 2015* (eds. Navab, N., Hornegger, J., Wells, W.M. & Frangi, A.F.) 234–241 (Springer International Publishing, 2015).
11. Naylor, P., La, M., Reyal, F. & Walter, T. Segmentation of nuclei in histopathology images by deep regression of the distance map. *IEEE Trans. Med. Imaging* **0062**, 1–12 (2018).
12. Simpson, J.C. et al. Genome-wide RNAi screening identifies human proteins with a regulatory function in the early secretory pathway. *Nat. Cell Biol.* **14**, 764–774 (2012).
13. Carpenter, A.E. et al. CellProfiler: Image analysis software for identifying and quantifying cell phenotypes. *Genome Biol.* **7**, R100 (2006).
14. Collinet, C. et al. Systems survey of endocytosis by multiparametric image analysis. *Nature* **464**, 243–249 (2010).
15. Graml, V. et al. A genomic multiprocess survey of machineries that control and link cell shape, microtubule organization, and cell-cycle progression. *Dev. Cell* **31**, 227–239 (2014).
16. Boyd, J., Pinhiero, A., Nery, E.D., Reyal, F. & Walter, T. Analysing double-strand breaks in cultured cells for drug screening applications by causal inference. In *International Symposium on Biomedical Imaging (ISBI): From Nano to Macro 2018–April*, 445–448 (2018).
17. Walter, T. et al. Automatic identification and clustering of chromosome phenotypes in a genome wide RNAi screen by time-lapse imaging. *J. Struct. Biol.* **170**, 1–9 (2010).
18. Neumann, B. et al. Phenotypic profiling of the human genome by time-lapse microscopy reveals cell division genes. *Nature* **464**, 721–727 (2010).
19. Dao, D. et al. CellProfiler analyst: Interactive data exploration, analysis and classification of large biological image sets. **32**, 3210–3212 (2016).
20. Piccinini, F. et al. Advanced cell classifier: User-friendly machine- learning-based software for discovering phenotypes in high-content imaging data. *Cell Syst.* **4**, 651.e5–655.e5 (2017).
21. Buggenthin, F. et al. Prospective identification of hematopoietic lineage choice by deep learning. *Nat. Methods* **14**, 403–406 (2017).

22. Caicedo, J.C. et al. Data-analysis strategies for image-based cell profiling. *Nat. Methods* **14**, 849–863 (2017).
23. Samacoits, A. et al. A computational framework to study sub-cellular RNA localization. *Nat. Commun.* **9**, 4584 (2018).
24. Coelho, L.P. et al. Determining the subcellular location of new proteins from microscope images using local features. *Bioinformatics* **29**, 2343–2349 (2013).
25. Boland, M.V., Markey, M.K. & Murphy, R.F. Automated recognition of patterns characteristic of subcellular structures in fluorescence microscopy images. *Cytometry* **33**, 366–375 (1998).
26. Murphy, R.F. & Wang, Y. Putting proteins on the map. *Nat. Biotechnol.* **24**, 1223–1224 (2006).
27. Glory, E. & Murphy, R.F. Automated subcellular location determination and high-throughput microscopy. *Dev. Cell* **12**, 7–16 (2007).
28. Schoenauer Sebag, A. et al. Infering an ontology of single cell motions from high-throughput microscopy data. In *12th International Symposium on Biomedical Imaging (ISBI)* 160–163 (2015). doi:10.1109/ISBI.2015.7163840.
29. Sullivan, D.P. et al. Deep learning is combined with massive-scale citizen science to improve large-scale image classification. *Nat. Biotechnol.* **36**, 820–828 (2018).
30. Christiansen, E.M. et al. In silico labeling: Predicting fluorescent labels in unlabeled images. *Cell* **173**, 792.e19–803.e19 (2018).

# Conclusions and Future Perspectives

T HE UNSOLVED CHALLENGE IN cancer research and in clinics is integrating together various and permanently evolving types of data. The second related persisting challenge is making them meaningful and accessible for researchers from different (non-computational) fields, to clinicians and finally to the patients themselves. Artificial intelligence (AI) approaches may provide the essential solutions here.

In the field of cancer genomics, the broad availability of genetic information provided by advanced sequencing technologies needs to be fast and efficiently analyzed using AI approaches and interpreted in the context of additional characteristics from clinical records or molecular knowledge summarized in pathway databases. However, the knowledge rapidly evolves as well, thus requiring advanced ML text mining mechanisms to systematically analyze all scientific papers and support the update molecular networks and pathway databases.

Cancer nowadays is a chronic disease, often evolving over decades, naturally making it coincide with other diseases. Advanced methods are needed to first retrieve the clinical information from the records, and thus associate with the multilevel omics data variables and further evaluate disease comorbidities. Obviously, multiple diseases in a patient will require very thoughtful treatment design, probably involving drug repositioning. To make such a decision, advanced ML algorithms need to be applied.

Further, one needs to transform big data into a clinically actionable source of information, e.g. evidence-based clinical recommendations. However, as clinical records become more accurate, they simultaneously become much more elaborate and loaded with data on each patient. Therefore, AI adoption in digital healthcare with regard to data management and accessibility is absolutely essential.

# List of Acronyms

## CHAPTER 1.1

**SIGNOR:** Signalling Network Open Resource
**DISNOR:** DISease Network Open Resource
**GDA:** gene-disease association
**KEGG:** Kyoto Encyclopedia of Genes and Genomes
**PSI-MI:** Proteomics Standards Initiative-Molecular Interactions

## CHAPTER 1.2

**BioPax:** Biological Pathway Exchange
**SBML:** Systems Biology Markup Language
**UniProt:** Universal Protein Resource

## CHAPTER 1.3

**ACSN:** Atlas of Cancer Signalling Network
**DNA:** deoxyribonucleic acid
**RNA:** ribonucleic acid
**TME:** tumour microenvironment
**CAF:** cancer associated fibroblasts

## CHAPTERS 2.1 AND 2.2

**IHC:** immunohistochemistry
**FFPE:** formalin-fixed, paraffin-embedded
**TCR:** T cell receptor
**Ig:** immunoglobulin

## CHAPTER 2.3

**CSC:** cancer stem cells
**TMEGA:** Tumour MicroEnvironment Global Analysis
**RNA:** ribonucleic acid

**scRNAseq:** single-cell RNA sequencing
**DC:** dendritic cells
**ICELLNET:** InterCELLular communication NETwork reconstruction

## CHAPTERS 3.1 AND CHAPTER 3.2

**COPD:** Chronic Obstructive Pulmonary Disease
**RMS:** relative molecular similarity
**SCN:** Stratified Comorbidity Network

## CHAPTERS 3.3, 4.1 AND 4.2

**EGFR:** epidermal growth factor receptor
**IκB:** inhibitor of κB binding
**IKK:** IκB kinase
**NF-κB:** nuclear factor-κB
**qPCR:** quantitative polymerase chain reaction
**RC3H1:** ring finger and CCCH-type domains 1
**SRM:** selected-reaction monitoring
**TNFα:** tumour necrosis factor-alpha
**TRAF:** TNF receptor associated factor

## CHAPTER 4.3

**PROGENy:** Pathway RespOnsive GENes
**DoRothEA:** Discriminant Regulon Expression Analysis
**CellNOpt:** Cellular Network Optimizer
**Carnival:** CAusal Reasoning for Network identification using Integer
         VALue programming

## CHAPTER 4.4

**MaBoSS:** Markovian Boolean Stochastic Simulator
**PROFILE:** PeRsonalization OF logIcaL ModEls

## CHAPTER 5.1

**HSC:** hematopoieitic stem cell
**MPP:** multipotent progenitor
**DNA:** deoxyribonucleic acid
**RNA:** ribonucleic acid
**PCA:** principal component analysis
**CRISPR:** clustered regularly interspaced short palindromic repeats
**Cas9:** CRISPR associated protein 9

## CHAPTERS 5.2, 6.1, 6.2 AND 6.3

**AUC:** area under curve
**IDH1:** Isocitrate dehydrogenase of type 1 gene
**MET:** Methionine
**MRI:** magnetic resonance imaging
**PDE:** partial differential equations
**PET:** positron emission tomography
**SUV:** standard uptake value

## CHAPTER 6.4

**LDA:** linear discriminant analysis
**SVM:** support vector machines
**RF:** random forests
**CNN:** convolutional neural networks
**PCA:** principal component analysis
**t-SNE:** t-Distributed Stochastic Neighbour Embedding
**smFISH:** single molecule in situ hybridization
**HCS:** high content screening

# Glossary

**Chapter 1.1**

## CAUSAL INTERACTIONS

Causal interactions involve two entities, one being the regulator, and the second one being the target of the regulation. The regulation can be positive or negative.

## DISGENET

A discovery platform for the dynamical exploration of human diseases and their genes.

## GRAPH

Diagram used to display the relationship between interacting entities. Entities are called nodes (or vertex) and are connected by edges.

## MANUAL CURATION

Manual curation is the process through which the information is translated from a human-readable format (in the literature) into a structured computer-readable format (in a database).

**Chapter 1.2**

## BENJAMINI-HOCHBERG METHOD

This method helps to reduce false positives by ranking and eliminating hypotheses.

## ENRICHED/OVER-REPRESENTED PATHWAYS

Pathways that have more proteins from the user input list than would be expected by chance.

## FALSE DISCOVERY RATE

The likelihood of an incorrect rejection of a hypothesis.

## HYPERGEOMETRIC DISTRIBUTION TEST

This test describes the probability of $k$ successes in $n$ draws without replacement from a finite population.

**Chapter 1.3**

## ACSN

Atlas of Cancer Signalling Network, an interactive and comprehensive map of molecular mechanisms implicated in cancer.

## CELLDESIGNER

A structured diagram editor for drawing gene-regulatory and biochemical networks.

## FAIR PRINCIPLES

A set of guiding principles in order to make data findable, accessible, interoperable and reusable. These principles provide guidance for scientific data management.

## GARUDA

A connectivity platform to connect, discover and navigate through interoperable gadgets with applications in healthcare and beyond.

## HALLMARKS OF CANCER

The hallmarks of cancer are a set of biological capabilities acquired during the multistep development of human tumours.

## MINERVA

Molecular Interaction NEtwoRk VisuAlization platform is a standalone webserver for visualization, exploration and management of molecular networks encoded in SBGN-compliant format.

## NAVICELL

A web-based environment for exploiting large maps of molecular interactions and data visualization.

## NDEX
Network Data Exchange is a collaborative software infrastructure for storing, sharing and publishing biological network knowledge.

## RECONMAP
An interactive visualization tool of human metabolism.

## VIRTUAL METABOLIC HUMAN
A database encapsulating current knowledge of human metabolism. Compiles information related to the links between human metabolism and genetics, microbial metabolism, nutrition and disease.

**Chapter 2.2**

## BAYESIAN CLUSTERING APPROACH
A method producing posterior distributions of all cluster parameters and proportions, in addition to associated cluster probabilities for all objects.

## MIXCR
A free and non-profit tool to process big immunome data from raw RNA or DNA sequences to quantitate clonotypes.

## PARSIMONY RATCHET METHOD
The Parsimony Ratchet is a method for rapid parsimony analysis of large datasets.

## TNM CLASSIFICATION
The Tumour Node Metastasis (TNM) staging system is an anatomically based classification of tumours currently used to guide treatment stratification. T represents the size and extent of the primary Tumour; N the involvement of regional lymph Nodes; and M the presence of distant Metastases.

**Chapter 2.3**

## ICELLNET
The InterCELLular communication NETwork reconstruction tool is a transcriptomic-based tool to reconstruct intercellular communication networks.

## TMEGA

The Tumour MicroEnvironment Global Analysis is an original program to analyze the global cellularity and heterogeneity in human primary tumours.

**Chapter 3.2**

## COMORBIDITY

Co-occurrence of one or more disorders in the same child or adolescent either at the same time or in some causal sequence.

**Chapter 5.1**

## FATEID

FateID is a method for the quantification of cell fate bias in single-cell transcriptome datasets comprising different cell types that emerge from a common progenitor.

## RACEID

RaceID is a method for rare cell type identification from single-cell RNA-seq data by unsupervised learning.

## STEMID

StemID is an algorithm for the inference of lineage trees and differentiation trajectories based on pseudo-temporal ordering of single-cell transcriptomes and utilizes the clusters predicted by RaceID.

## STEMNET

STEMNET is method for the reconstruction of stem cell lineage priming from single-cell data.

**Chapter 6.3**

## IDH1

Isocitrate dehydrogenase of type 1 gene is involved in gliomas. Patients with the mutated gene seem to have a better clinical outcome.

**Chapter 6.4**

## DEEP LEARNING

Powerful learning technique allowing the ability to perform classification without prior definition of object features.

## HIGH CONTENT SCREENING

Large-scale imaging approaches consisting of thousands of experiments relying on automated microscopy.

## SINGLE MOLECULE FISH (SMFISH)

Technique to visualize individual RNA molecules inside cells and/or organisms.

## SUPERVISED LEARNING

Learning of classification rules from annotated samples.

## UNSUPERVISED LEARNING

Discovering patterns (clusters) in datasets without the use of annotated samples.

# Index

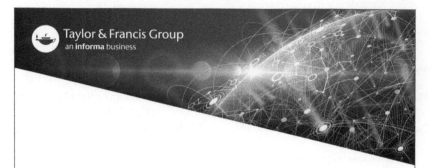